Sugarcane ethanol

Sugarcane ethanol

Contributions to climate change mitigation and the environment

edited by:

Peter Zuurbier

Jos van de Vooren

ISBN 978-90-8686-090-6

First published, 2008

Wageningen Academic Publishers
The Netherlands, 2008

Table of contents

Foreword

José Goldemberg, professor at the University of São Paulo, Brazil

Ethanol, produced from biomass, has been considered as a suitable automobile fuel since the beginning of the automotive industry one century ago, particularly for vehicles powered with spark-ignition engines (technically referred as Otto cycle engines, but commonly known as gasoline engines). However, the use of ethanol was dwarfed by gasoline refined from abundant and cheap oil. The staggering amounts of gasoline in use today – more than 1 trillion litres per year – eliminated almost all the alternatives.

However environmental as well as security of supply concerns sparked, in the last decades, renewed interest in ethanol. In many countries it is blended with gasoline in small amounts to replace MTBE. In Brazil it has already replaced 50% of the gasoline thanks to the use of flex-fuel engines or dedicated pure ethanol motors. Worldwide ethanol is replacing already 3% of the gasoline.

Maize (in the US) and sugarcane (in Brazil) account for 80% of all ethanol in use today. The agricultural area used for that purpose amounts to 10 million hectares less than 1% of the arable land in use in the world.

There are three main routes to produce ethanol from biomass:

- fermentation of sugar from sugarcane, sugar beet and sorghum;
- saccharification of starch from maize, wheat and manioc;
- hydrolysis of cellulosic materials, still in development.

There are important differences between the fermentation and saccharification routes. When using sugarcane one does not need an 'external' source of energy for the industrial phase of ethanol production since the bagasse supplies all the energy needed. The fossil fuel inputs are small (in the form of fertilizers, pesticides, etc.) so basically this route converts solar energy into ethanol. The final product is practically a renewable fuel contributing little to greenhouse gas (GHG) emissions.

Ethanol from maize and other feed stocks requires considerable inputs of 'external' energy most of it coming from fossil fuels reducing only marginally GHG emissions.

Sugarcane grows only in tropical areas and the Brazilian experience in this area led to ethanol produced at very low cost and competitive with gasoline through gains in productivity and economies of scale (Goldemberg, 2007). Ethanol produced from maize in the US cost almost twice and from wheat, sugar beets, sorghum (mainly in Europe) four times (Worldwatch Institute, 2006).

The use of biofuels as a substitute for gasoline has been recently criticized mainly for:

- sparking a competition between the use of land for fuel 'versus' land for food which is causing famine in the world and
- leading to deforestation in the Amazonia.

The importance of these concerns was greatly exaggerated and is, generally speaking, unwarranted.

The recent rise in prices of agricultural products – after several decades of declining real prices – has given rise to the politically laden controversy of fuel 'versus' food. This problem has been extensively analyzed in many reports, particularly the World Bank (World Bank, 2008), which pointed out that grain prices have risen due to a number of individual factors, whose combined effect has led to an upward price spiral namely: high energy and fertilizer prices, the continuing depreciation of the US dollar, drought in Australia, growing global demand for grains (particularly in China), changes in import-export policies of some countries and speculative activity on future commodities trading and regional problems driven by policies subsidizing production of biofuels in the US and Europe (from maize, sugar beets and wheat). The expansion of biofuels production particularly from maize over areas covered by soybeans in the US contributed to price increases but was not the dominant factor. The production of ethanol from sugarcane in Brazil has not influenced the prize of sugar.

Despite that, the point has been made that other countries had to expand soybean production to compensate for reductions in the US production possibly in the Amazonia, increasing thus deforestation. Such speculative 'domino effect' is not borne out by the facts: the area used for soybeans in Brazil (mainly in the Amazonia) has not increased since 2004 (Goldemberg and Guardabassi, in press). The reality is that deforestation in the Amazonia has been going on for a long time at a rate of approximately 1 million hectares per year and recent increases are not due to soybean expansion but to cattle.

Emissions from land use changes resulting from massive deforestation would of course release large amounts of CO_2 but the expansion of the sugarcane plantations in Brazil is taking place over degraded pastures very far from the Amazonia. Emissions from such land use change have been shown to be small (Cerri *et al.*, 2007).

The present area used of sugarcane for ethanol production in Brazil today is approximately 4 million hectares out of 20 million hectares used in the world by sugarcane in almost 100 countries. Increasing the areas used for of sugarcane for ethanol production in these countries by 10 million hectares would result in enough ethanol to replace 10% of the gasoline in the world leading to a reduction of approximately 50 million tons of carbon per year. This would help significantly many OECD countries to meet the policy mandates adopted for the use of biofuels.

Such course of action would of course require a balanced weighting of the advantages of replacing gasoline by a renewable fuel and impacts and land use and biodiversity.

This book analyzes all these aspects of the problem and will certainly be an important instrument to clarify the issues, dispel some myths and evaluate the consequence of different policy choices.

References

Cerri, C.E.P., M. Easter, K. Paustian, K. Killian, K. Coleman, M. Bernoux, P. Falloon, D.S. Powlson, N.H. Batjes, E. Milne and C.C. Cerri, 2007. Predicted soil organic carbon stocks and changes in the Brazilian Amazon between 2000 and 2030. Agriculture, Ecosystems and Environment 122: 58-72.

Goldemberg, J., 2007. Ethanol for a Sustainable Energy Future. Science 315: 808-810.

Goldemberg, J., S.T. Coelho and P. Guardabassi, 2008. The sustainablility of ethanol production from sugarcane. Energy Policy 36: 2086-2097.

Worldwatch Institute, 2006. Biofuels for transport: Global Potential and Implications for Sustainable Agriculture and Energy in 21st Century. ISBN 978-1-84407-422-8

World Bank, 2008. Double Jeopardy: Responding to high Food and Fuel Prices. G8 Hokkaido – Toyako Summit. July 2, 2008.

Executive summary

Do biofuels help to reduce greenhouse gas emissions and do they offer new sources of income to farmers, by producing biomass? Are biofuels competing with food, animal feed and contributing to higher food prices? And are biofuels directly or indirectly threatening the environment, biodiversity, causing irreversible or undesirable changes in land use and landscape?

This publication aims to set the stage for the discussion about both challenges and concerns of sugarcane ethanol by providing the scientific context, the basic concepts and the approach for understanding the debate on biofuel-related issues. This book largely limits itself to sugarcane ethanol and its contribution to climate change mitigation and the environment.

The main findings and conclusions are:

1. The dominance of Brazil in global sugarcane production and expansion – Brazil accounted for 75 percent of sugarcane area increase in the period 2000 to 2007 and two-thirds of global production increase in that period – derives from its experience and capability to respond to thriving demand for transport fuels, which was recently triggered by measures to mitigate greenhouse gas emissions of the rapidly growing transport sector, concerns in developed countries to enhance energy security and lessen dependence on petroleum, and not the least the need of many developing countries to reduce import bills for fossil oil.
2. According to the IIASA/AEZ assessment, the most suitable climates for rain-fed sugarcane production are found in south-eastern parts of South America, e.g. including São Paulo State in Brazil, but also large areas in Central Africa as well as some areas in Southeast Asia. The massive further expansion of sugarcane areas, e.g. as forecasted for Brazil, is expected to cause the conversion of pastoral lands in the savannah region.
3. This study analyzes the land use changes (LUC) in Brazil caused by sugarcane expansion, looking both at the past and expected future dynamics. Remote sensing images have identified that in 2007 and 2008 Pasture and Agriculture classes together were responsible for almost 99% of the total area displaced for sugarcane expansion which equals an area of more than 2 million ha. Pasture was responsible for approximately 45% and Agriculture was responsible for more than 50% of the displaced area for sugarcane. About 1% of sugarcane expansion took place over the Citrus class and less than 1% over the Reforestation and Forest classes together. Pasture displacement is more important in São Paulo and Mato Grosso do Sul, while Agriculture is more important in the other states analyzed.
4. The shift-share model using IBGE micro-regional data has analyzed sugarcane expansion from 2002 to 2006 and has identified around 1 million ha in the ten Brazilian states analyzed. From this total expansion, 773 thousand ha displaced pasture land and 103 thousands displaced other crops, while only 125 thousand ha were not able to be allocated

over previous productive areas (meaning new land has been incorporated into agricultural production, which might be attributed to the conversion of forest to agriculture or to the use of previously idle areas). Total agricultural area growth – the sum of all crops, including sugarcane, and pastures – in the period was around 3.3 million ha.

5. Projections indicate that harvested sugarcane area in Brazil will reach 11.7 million ha and other crops 43.8 million ha in 2018, while pasture area will decrease around 3 million ha. The total land area in Brazil is 851.196.500 ha.
6. The expansion of crops, except sugarcane, and pasture land is taking place despite of the sugarcane expansion. This is important because it reinforces that, even recognizing that sugarcane expansion contributes to the displacement of other crops and pasture, there is no evidence that deforestation caused by indirect land use effect is a consequence of sugarcane expansion.
7. Sugarcane ethanol from Brazil does comply with the targets of greenhouse gases (GHG) reduction.
8. The GHG emissions and mitigation from fuel ethanol production/use in Brazil are evaluated for the 2006/07 season, and for two scenarios for 2020: the 2020 Electricity Scenario (already being implemented) aiming at increasing electricity surplus with cane biomass residues; and the 2020 Ethanol Scenario using the residues for ethanol production. Emissions are evaluated from cane production to ethanol end use; process data was obtained from 40 mills in Brazilian Centre South. Energy ratios grow from 9.4 (2006) to 12.1 (2020, the two Scenarios); and the corresponding GHG mitigation increase from 79% (2006) to 86% (2020) if only the ethanol is considered. With co-products (electricity) it would be 120%. LUC derived GHG emissions were negative in the period 2002 – 2008, and very little impact (if any) is expected for 2008 – 2020, due mostly to the large availability of land with poor carbon stocks. Although indirect land use changes (ILUC) impacts cannot be adequately evaluated today, specific conditions in Brazil may lead to significant increases in ethanol production without positive ILUC emissions.
9. Brazil has achieved very high levels of productivity (on average 7.000 litres of ethanol/ha and 6,1 MWhr of energy/ha), despite its lower inputs of fertilizers and agrochemicals compared with other biofuels, while reducing significantly the emissions of greenhouse gases. The ending of sugarcane burning in 2014 is a good example of improving existing practices.
10. Production of ethanol in Brazil, which has been rising fast, is expected to reach 70 billion litres by the end of 2008. Approximately 80% of this volume will be used in the transport sector while the rest will go into alcoholic beverages or will be either used for industrial purposes (solvent, disinfectant, chemical feedstock, etc.).
11. When evaluating key drivers for ethanol demand, energy security and climate change are considered to be the most important objectives reported by nearly all countries that engage in bioenergy development activities. A next factor is the growth in demand for transport fuels. A third factor is vehicle technologies that already enable large scale use of ethanol.

12. Projections of ethanol production for Brazil, the USA and the EU indicate that supply of 165 billion litres by 2020 could be achieved with the use of a combination of first and second generation ethanol production technologies.
13. Compared to current average vehicle performance, considerable improvements are possible in drive chain technologies and their respective efficiencies and emission profiles. IEA does project that in a timeframe towards 2030, increased vehicle efficiency will play a significant role in slowing down the growth in demand for transport fuels. With further technology refinements, which could include direct injection and regenerative breaking, fuel ethanol economy of 24 km/litre may be possible. Such operating conditions, can also deliver very low emissions.
14. Future ethanol markets could be characterized by a diverse set of supplying and producing regions. From the current fairly concentrated supply (and demand) of ethanol, a future international market could evolve into a truly global market, supplied by many producers, resulting in stable and reliable biofuel sources. This balancing role of an open market and trade is a crucial precondition for developing ethanol production capacities worldwide.
15. However, the combination of lignocellulosic resources (biomass residues on shorter term and cultivated biomass on medium term) and second generation conversion technology offers a very strong perspective. Also, the economic perspectives for such second generation concepts are very strong, offering competitiveness with oil prices equivalent to some 55 US$/barrel around 2020.
16. First generation biofuels in temperate regions (EU, North America) do not offer a sustainable possibility in the long term: they remain expensive compared to gasoline and diesel (even at high oil prices), are often inefficient in terms of net energy and GHG gains and have a less desirable environmental impact. Furthermore, they can only be produced on higher quality farmland in direct competition with food production. Sugarcane based ethanol production and to a certain extent palm oil and Jatropha oilseeds are notable exceptions to this, given their high production efficiencies and low(er) costs.
17. Especially promising are the production via advanced conversion concepts biomass-derived fuels such as methanol, hydrogen, and ethanol from lignocellulosic biomass. Ethanol produced from sugarcane is already a competitive biofuel in tropical regions and further improvements are possible. Both hydrolysis-based ethanol production and production of synthetic fuels via advanced gasification from biomass of around 2 Euro/GJ can deliver high quality fuels at a competitive price with oil down to US$55/barrel. Net energy yields per unit of land surface are high and up to a 90% reduction in GHG emissions can be achieved. This requires a development and commercialization pathway of 10-20 years, depending very much on targeted and stable policy support and frameworks.
18. Global land use changes induced by US and EU biofuels mandates show that when it comes to the assessing the impacts of these mandates on third economies, the combined policies have a much greater impact than just the US or just the EU policies alone, with crop cover rising sharply in Latin America, Africa and Oceania as a result of the biofuel

mandates. These increases in crop cover come at the expense of pasturelands (first and foremost) as well as commercial forests.

19. Sugarcane based ethanol can contribute to the achievement of several Millennium Development Goals through a varied range of environmental, social and economic advantages over fossil fuels. These include enhanced energy security both at national and local level; improved trade balance by reducing oil imports; improved social well-being through better energy services especially among the poorest; promotion of rural development and better livelihoods; product diversification leaving countries better-off to deal with market fluctuations; the creation of new exports opportunities; the potential to help tackling climate change through reduced emissions of greenhouse gases as well as other air emissions; and opportunities for investment attraction through the carbon finance markets. The highest impact on poverty reduction is likely to occur where sugarcane ethanol production focuses on local consumption, involving the participation and ownership of small farmers and where processing facilities are near to the cultivation fields.
20. Development of oil prices is crucial for the development of biofuels. High feedstock prices make biofuels less profitable. Hence, price hikes for commodities have a negative impact on bioethanol prices. Other factors, like stock level, price speculation, expected policy measures and natural disasters may add to price volatility as well.

The final conclusion is that sugarcane ethanol contributes to mitigation of climate change. The environmental impacts of sugarcane ethanol production are overall positive within certain conditions, as outlined in this publication, For advancing the sustainable sugarcane ethanol production, it is of importance to enhance a process of dialogue in the market place and between interested stakeholders in society.

Chapter 1
Introduction to sugarcane ethanol contributions to climate change mitigation and the environment

Peter Zuurbier and Jos van de Vooren

1. Introduction

Life is energy. Humankind depends on energy and produces and consumes large volumes of energy. The total final energy consumption in industry, households, services and transport in 2005 was 285 EJ (OECD/IEA, 2008). And the consumption is growing fast. The growth of global final energy between 1990 and 2005 was 23%. Globally, energy consumption grew most quickly in the transport and service sectors. Between 1990 and 2005, global final energy use in transport increased by 37% to 75 EJ and according to the IEA study, road transport contributes the most to the increase in overall transport energy consumption. Between 1990 and 2005, road transport energy use increased by 41%. And with this growth, CO_2 emissions increased as well. These emissions grew during that same period with 25% (IEA, 2008). The associated CO_2 emissions increased to 5.3 Gt CO_2. There is a widely shared opinion that these emissions contribute to global warming and climate change. Reason enough for making a change.

Another reason for making a change, are the fossil oil prices. Fact is that the price increased from $20 in 2002 to a record high of more than $140 a barrel in July 2008. The price volatility creates a lot of uncertainty in global markets. So, it is not surprising that the world is looking for substitutes for petroleum-derived products. Securing a reliable, constant and sustainable supply of energy demands a diversification of energy sources and an efficient use of available energy.

One of the alternatives for fossil fuels is biofuels. And here we enter in to the heat of the debate. Do biofuels help to reduce greenhouse gas emissions and offering new sources of income to farmers, by producing biomass? Are biofuels competing with food, animal feed and contributing to higher food prices? And are biofuels directly or indirectly threatening the environment, biodiversity, causing irreversible or undesirable changes in land use and landscape?

In this publication we aim to set the stage for the discussion about both challenges and concerns of sugarcane ethanol by providing the scientific context, the basic concepts and the approach for understanding the debate on biofuel-related issues. This book largely limits itself to sugarcane ethanol and its contribution to climate change mitigation and the environment.

2. Biofuels

Biofuels encompass a variety of feedstock, conversion technologies, and end uses. They are used mostly for transport and producing electricity. Biofuels for transportation, like ethanol and biodiesel, are one of the fastest-growing sources of alternative energy in the world today. Global production of biofuels amounted to 62 billion litres or 36 million tonnes of oil equivalent (Mt) in 2007 - equal to about 2 % of total global transport fuel consumption in energy terms (OESO, 2008).

3. Bioethanol

Global bioethanol production tripled from its 2000 level and reached 52 billion litres (28.6 Mt) in 2007 (OESO, 2008). Based on the origin of supply, Brazilian ethanol from sugarcane and American ethanol from maize are by far leading the ethanol production. In 2007 Brazil and the United States together accounted for almost 90% of the world ethanol production.

In Brazil production of ethanol, entirely based on sugarcane (*Saccharum* spp.), started in the seventies and peaked in the 1980s, then declined as international fossil oil prices fell back, but increased rapidly again since the beginning of the 21st century. Falling production costs, higher oil prices and the introduction of vehicles that allow switching between ethanol and conventional gasoline have led to this renewed surge in output.

In the crop season 2007/08 Brazil produced 22.24 billion litres of ethanol. Conab/AgraFNP expects another jump for the crop season 2008/09 with an expected production of 26.7 billion litres (AgraFNP, 2008). This increase is mainly due to expansion of the sugarcane area. In 2007/08 the area for sugarcane was 6.96 million hectare, and is estimated to grow to 7.67 million hectare in 2008/09. The total sugarcane production will also increase from 549.902 Mt to 598.224 Mt.

A typical plant in Brazil crushes 2 million tonnes of sugarcane per year and produces 200 million litres of ethanol per year (1 million litres per day during 6 months – April to November in the south-eastern region). The size of the planted area required to supply the processing plant is on average 30,000 hectares. Due to process of degradation of the quality of harvested cane the distance to the mill is up to 70 kilometres at the most.

United States (US) output of ethanol, mainly from maize (*Zea mays* ssp. *mays* L.), has increased in recent years as a result of public policies and measures such as tax incentives and mandates and a demand for ethanol as a replacement for methyl-tertiary-butyl-ether (MTBE) a gasoline-blending component. Between 2001 and 2007, US fuel ethanol production capacity grew 220 from 7.19 billion to 26.50 billion litres (OECD, 2008). The new Energy Bill expands the mandate for biofuels, such as ethanol, to 56.8 billion litres in 2015.

Although the installed ethanol fuel capacity in the European Union (EU) amounts to 4.04 billion litres at the moment (OESO, 2008), Europe's operational capacity is significantly lower at 2.9-3.2 billion litres as some plants have suspended production. The bulk of EU production, however, is biodiesel, which, in turn, accounts for almost two-thirds of world biodiesel output.

Elsewhere, China with 1.8 billion litres of ethanol (Latner *et al.*, 2007), Canada with 0.8 billion litres are relatively smaller producers.

4. Production and use of bioethanol

Ethanol is manufactured by microbial conversion of biomass materials through fermentation. The production process consists of three main stages:

- conversion of biomass to fermentable sugars;
- fermentation of sugars to ethanol; and
- separation and purification of the ethanol (Figure 1).

Fermentation initially produces ethanol containing a substantial amount of water. Distillation removes the major part of the water to yield about 95 percent pure ethanol. This mixture of 95% ethanol and water is called hydrous ethanol. If the remaining water is removed, the ethanol is called anhydrous ethanol and is suitable for blending with gasoline. Ethanol is 'denatured' prior to leaving the distillery to make it unfit for human consumption.

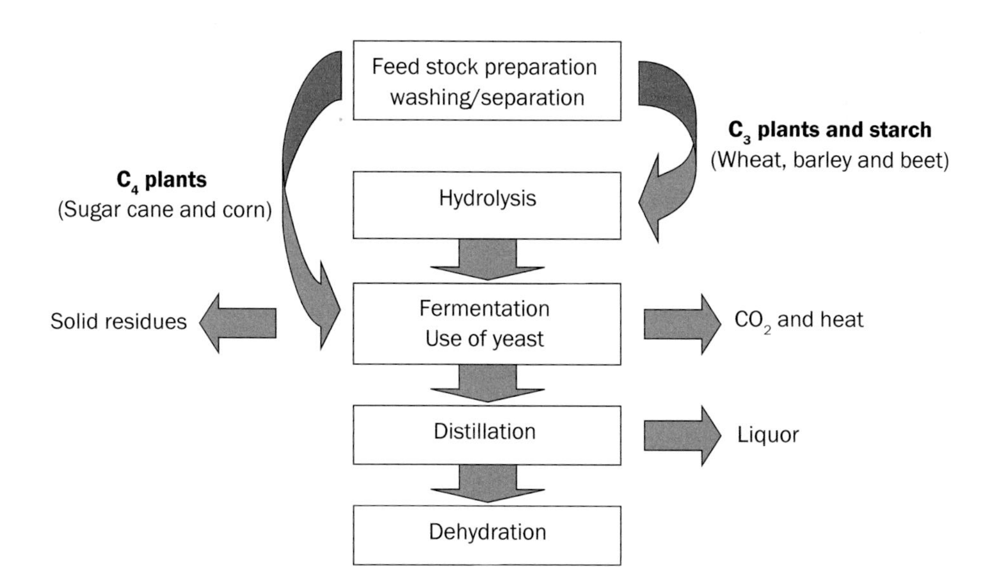

Figure 1. Production process of ethanol (Barriga, 2003).

Traditional fermentation processes rely on yeasts that convert six-carbon sugars, such as glucose, into ethanol. Ethanol is used primarily in spark-ignition engine vehicles. The amount of ethanol in the fuel ranges from 100 percent to 5 percent or lower, blended with gasoline. In Brazil the Flex-Fuel-Vehicles (FFV) are fit to use the whole range of blends of ethanol, up to 100%. The attractiveness of FFV is shown by the fact that in 2008 of the new cars sold 87.6% are FFV's (Anfavea: www.anfavea.com.br/tabelas.html). In other countries, such as Sweden, a maximum of 85% (E85) is used.

Anhydrous ethanol is used in a gasoline-ethanol blend. For example, of the total Brazilian ethanol production in the crop-season 2007/08, 8.38 billion litres are anhydrous and the rest, 13.86 billion litres hydrous ethanol (AgraFNP, 2008). Aside from FFV's manufactured to run on hydrous ethanol, non-FFV's in Brazil run on a 25 % mixture of a gasoline-ethanol blend and hydrous ethanol.

Another application of ethanol is as a feedstock to make ethers, most commonly ethyl tertiary-butyl ether (ETBE), an oxygenate with high blending octane used in gasoline. ETBE contains 44 percent ethanol. A last application, that we mention here, is the use of ethanol in diesel engines. Take for example Scania: Scania's compression-ignition (CI) ethanol engine is a modified 9-liter diesel with a few modifications. Scania raised the compression ratio from 18:1 to 28:1, added larger fuel injection nozzles, and altered the injection timing. The fuel system also needs different gaskets and filters, and a larger fuel tank since the engine burns 65% to 70% more ethanol than diesel. The thermal efficiency of the engine is comparable to a diesel, 43% compared to 44% (http://gas2.org/2008/04/15).

5. Where does it come from: the feedstock for ethanol

The term feedstock refers to the raw material used in the conversion process. The main types of feedstock for ethanol are described below.

1. *Sugar and starch-based crops*: As mentioned earlier bioethanol is mainly produced of sugarcane and maize. Other major crops being used are wheat, sugar beet, sorghum and cassava. Starch consists of long chains of glucose molecules. Hydrolysis, a reaction of starch with water, breaks down the starch into fermentable sugars (see Figure 1).
   The co-products include bagasse (the residual woody fibre of the cane obtained after crushing cane), which can be used for heat and power generation in the case of sugarcane; distiller's dried grains sold as an animal feed supplement from maize in dry mill processing plants; and high-fructose maize syrup, dextrose, glucose syrup, vitamins, food and feed additives, maize gluten meal, maize gluten feed, maize germ meal and maize oil in wet mill processing plants. In all cases, commercial carbon dioxide (CO_2) can be captured for sale.
2. *Wastes, residues and cellulosic material*: according to Kim and Dale (2005), there are about 73.9 million tonnes of dry wasted crops and about 1.5 billion tonnes of dry lignocellulosic biomass.

Cellulose is the substance that makes up the cell walls of plant matter along with hemicellulose and lignin. Cellulose conversion technologies will allow the utilization of nongrain parts of crops like maize stover, rice husk, straws, sorghum stalk, bagasse from sugarcane and wood and wood residues. Among the cellulosic crops perennial grasses like switchgrass (*Panicum virgatum* L.) and Miscanthus are two crops considered to hold enormous potential for ethanol production. Perennial crops offer other advantages like lower rates of soil erosion and higher soil carbon sequestration (Khanna *et al.*, 2007; Schuman et al, 2002) However, technologies for conversion of cellulose to ethanol are just emerging and not yet technically or commercially mature.
Furthermore, lignin-rich fermentation residue, which is the co-product of ethanol made from crop residues and sugarcane bagasse, can potentially generate electricity and steam.

6. Brazil as main exporter

Brazil has been by far the largest exporter of ethanol in recent years. In the crop season 2007/08, its hydrated ethanol exports amounted to 3.7 billion litres, of the 5 billion litres of ethanol traded globally (excl. intra-EU trade) (AgraFNP, 2008). The US imported more than half the ethanol traded in 2006. Of the 2.7 billion litres imported by the US in 2006, about 1.7 billion litres were imported directly from Brazil, while much of the remainder was imported from countries which are members of the Caribbean Basin Initiative (CBI) which enjoy preferential access to the US market and import (hydrated) ethanol from Brazil, dehydrate it and re-export to the US.

China, too, has been a net exporter of ethanol over the last several years, though at significantly lower levels than Brazil. Despite some exports to the US as well as to CBI countries, most of the larger destinations for Chinese ethanol are within the Asian region, in particular South Korea and Japan (OESO, 2008). The EU is also a net importer.

7. What makes the ethanol attractive?

One may observe a variety of reasons for the recent bioethanol interest. From the market point of view, there is an increasing consensus about the end of cheap oil and the volatility in world oil prices. Nowhere is the need for alternative to fossil oil felt more than in the transport sector. Transport consumes 30% of the global energy, 98 % of which is supplied by fossil oils (IEA, 2007).

From a policy point of view, other factors are mentioned, such as assuring energy security, reducing greenhouse gas emissions, increase and diversification of incomes of farmers and rural communities and rural development. And next there are arguments that ethanol is replenishable, that the ethanol industry can create new jobs, and that feedstock for ethanol can be made easily available considering already existing technologies.

However the debate on biofuels in general and bioethanol in particular shows a lot of counterarguments. They include that production of feedstock for ethanol might have negative environmental impacts on GHG, land use change, water consumption, biodiversity and air quality; also indirect negative environmental impacts are mentioned as a result of the interactions between different land uses. The development of biofuels, it is said, may also have both direct and indirect negative social and socio-economic impacts.

A third point of view comes from developing countries being motivated to diversify energy sources. Specifically net importing countries, may consider enhancing their energy security by domestically produced ethanol. Quality of air might be another argument for countries where the vehicle fleet is old, causing huge polluting emissions. However, also for these countries the counterarguments are widely discussed. Will the bioethanol production contribute to small farmers? And what will be the impact of production for bioethanol on the food production in those countries. Next to possible environmental impacts, developing countries might decide to take irreversible decisions that might, according to this point of view, create more instead off less poverty (Oxfam, 2008).

8. The core of the debate

The debate on sugarcane ethanol contains several major issues. The first one is impact of sugarcane production on land use change and climate. Here the assumption is made that land use for sugarcane implies serious impacts on the carbon stock, GHG emissions, and water and soil conditions. (Macedo *et al.*, 2004). Also, the reallocation of land or land cleared for ethanol may have unforeseen impacts on biodiversity. The main question here is, can production of sugarcane ethanol be sustainable?

Second, the demand side of the sugarcane ethanol may have impacts on the automotive industry, as happened in Brazil by the introduction of FFV's. Here the assumption is that demand will not so much be geared by balanced growth of the supply, but by the price and attractiveness of new automotive solutions. And this may have unintended consequences for sustainable production of sugarcane ethanol (Von Braun, 2006).

Third issue is the impact of new technologies on the efficiency of biomass for biofuels and the conversion of biomass for ethanol. Here the assumption is that new technologies may provide not only higher efficiency, but also the need for larger scale of operations, asking more land to be cleared for ethanol with possible negative environmental effects (Faaij, 2006).

Fourth, the public policies may have positive effects on balanced growth of the ethanol industry. However, these policies may also contribute to numerous distortions in trade, consumption, supply and technology development and on the environment as well (Hertel *et al.*, 2008).

Fifth, the debate also addresses the impacts of biofuels on developing countries. These societies may benefit greatly by diversifying the energy matrix. However, unbalanced growth may have unintended consequences for the food security domestically and land use (Teixeira Coelho, 2005; Kojima and Johnson, 2005; Dufey *et al.*, 2007).

Sixth, the last issue deals with the food prices hike. How do biofuels rank as factor for explaining the food prices in 2007-2008 and, possibly, the coming years (Banse, 2008; Maros and Martin, 2008)? And how does ethanol fit into this explanation and projection?

The impact studies are conducted from a multidisciplinary point of view. Also, the impacts are observed on different scale levels: global, regional and on value chain level. Hence, the analysis focuses on land use dynamics, market demand, technology development and public policies. These four main factors are assumed to contribute to the understanding of impacts of sugarcane ethanol on climate change mitigation and the environment. The debate asks understanding based on the latest science based insights (The Royal Society, 2008). This book aims to contribute to present these insights.

9. Structure of the book

In Chapter 2 the debate on sugarcane ethanol focuses on land use from a global point of view. There are many competing demands for land: to grow crops for food, feed, fibre and fuel, for nature conservation, urban development and other functions. The objective of the chapter is to analyze current and potential sugarcane production in the world and to provide an assessment of land suitable for sugarcane production.

Considering the particular situation in Brazil, Chapter 3 discusses the prospects of the sugarcane production, considering land use allocation and the land use dynamics. It shows on an empirical basis the expected sugarcane land expansion. This expansion is supposed to convert annual crops, permanent crops, pasture areas, natural vegetation and degraded areas. The chapter presents substitution patterns based on a reference scenario for sugarcane and ethanol production.

What are the impacts of sugarcane ethanol for the mitigation of GHG emissions? Chapter 4 goes into this debate. The chapter compares the ethanol production in 2006 with a scenario for 2020. Next energy flows and a life cycle analysis is presented. Then the effects on land use change on GHG emissions on global scale are discussed. Finally the chapter discusses the indirect effects of land use change in the Brazil.

Chapter 5 addresses the question on environmental sustainability of the sugarcane ethanol production in Brazil. Sustainable production is discussed worldwide. For bioethanol sustainability criteria vary among countries and institutions. Criteria that are pertinent in the debate are use of agricultural inputs, air quality and burning of sugarcane vs. mechanization, use of water, soil, farm inputs such as fertilizer and the energy and carbon balance. The chapter ends with the discussion on certification and compliance

Chapter 6 starts with the assessment of studies on the market potential of ethanol. The demand predictions will be considered, taking into consideration technological development and innovation. The study provides an overview of the main issues and challenges related to the current and potential use of ethanol in the transport sector.

In Chapter 7 the technology developments for bioenergy will be analyzed. It gives a state of the art overview of technologies for bioenergy production from biomass. Next the chapter highlights some challenges in developing technologies from biomass. Further it sheds light on some scenarios for technologies to be developed in the 10-15 years to come.

As described earlier, public policies play a major role in the biofuel industry. What are the policies, what measures are implemented and what are the impacts? This Chapter 8 will deal specifically with the policies originating from the United States of America and the European Union. The chapter starts with an overview of policies and policy instruments of both. Next, these policies will be evaluated from an economic point of view. Based on this analysis, the impacts on the global biofuel industry will be considered.

There is much debate on the impacts of biofuels on developing countries. Just positive, only negative? In Chapter 9 the impact will be discussed within the framework of the Millennium Development Goals (MDG). The chapter will deal with the question: How can global bio-fuels industry support sustainable development and poverty reduction?

The book ends with the probably most heated debate: the impacts of bio fuels production on food prices. Chapter 10 covers the following questions: what is the state of the art: what are the relations between production of food and food prices and bio-fuels? Then the main drivers for the hike in food prices are discussed. Based on quantitative model studies some core findings will be presented. Finally, the chapter ends with the impacts of bioethanol on food production and prices.

References

AgraFNP, 2008. June 24. Ethanol consumption and exports continue to increase.

Banse, M., P. Nowicki and H. van Meijl, 2008. Why are current world food prices so high? A memo. LEI Wageningen UR, The Hague, the Netherlands.

Barriga, A., 2003. Energy System II. University of Calgary/OLADE, Quito.

Dufey, A., S. Vermeulen and B. Vorley, 2007. Biofuels: Strategic Choices for Commodity Dependent Developing Countries. Common Fund for Commodities Amsterdam, the Netherlands.

Faaij, A., 2006. Modern Biomass Conversion Technologies. Mitigation and Adaptation Strategies for Global Change 11: 335-367.

Hertel, T., W. W.E. Tyner and D.K. Birur, 2008. Biofuels for all? Understanding the Global Impacts of Multinational. Center for Global Trade Analysis Department of Agricultural Economics, Purdue University GTAP Working Paper No. 51, 2008.
IEA, 2007. Bioenergy Potential contribution of bioenergy to the world's future energy demand, International Energy Agency, Paris.
IEA, 2008. Worldwide Trends in Energy Use and Efficiency Key Insights from IEA Indicator Analysis, Paris, France.
Khanna, M., H. Onal, B. Dhungana and M. Wander, 2007. Economics of Soil Carbon Sequestration Through Biomass Crops. Association of Environmental and Resource Economists; Workshop Valuation and Incentives for Ecosystem Services, June 7-9, 2007.
Kim, S. and B.E. Dale, 2005. Life cycle assessment of various cropping systems utilized for producing biofuels: Bioethanol and biodiesel. Biomass and Bioenergy 29: 426-439.
Kojima, M. and T. Johnson, 2005. Potential for biofuels for transport in developing countries. ESMAP, World Bank Copyright The International Bank for Reconstruction and Development/The World Bank, Washington D.C., USA.
Latner, K., O. Wagner and J. Junyang, 2007. China, Peoples Republic of Bio-Fuels Annual 2007. GAIN Report Number: CH7039. USDA Foreign Agricultural Service, January 2007.
Macedo, I.C., M.R.L.V. Leal and J.E.A.R. da Silva, 2004. Assessment of Greenhouse Gas Emissions in the Production and Use of Fuel Ethanol in Brazil. Report to the Government of the State of São Paulo, 2004.
Maros, I. and W. Martin, 2008. Implications of Higher Global Food Prices for Poverty in Low-Income Countries. The World Bank Development Research Group Trade Team April, Washington, USA.
OECD, 2008. Economic assessment of biofuel support policies. Paris, France.
OECD/IEA, 2008. Worldwide Trends in Energy Use and Efficiency Key Insights from IEA Indicator Analysis. Paris, France.
OESO, 2008. Economic assessment of biofuel support policies. Paris, France.
Oxfam, 2008. Inconvenient Truth How biofuel policies are deepening poverty and accelerating climate change Oxfam Briefing Paper, June 2008.
Schuman, G.E., H.H. Janzen and J.E. Herrick, 2002. Soil carbon dynamics and potential carbon sequestration by rangelands. Environmental Pollution 116: 391-396.
Teixeira Coelho, S., 2005. Biofuels- advantages and trade barriers. UNCTAD/DITC/TED/2005/1.
The Royal Society, 2008. Sustainable biofuels: prospects and challenges. London, United Kingdom. ISBN 9780854036622.
Von Braun, J., 2006. When Food Makes Fuel: The Promises and Challenges of Biofuels. Ifpri. Washington, USA.

Chapter 2
Land use dynamics and sugarcane production

Günther Fischer, Edmar Teixeira, Eva Tothne Hizsnyik and Harrij van Velthuizen

1. Historical scale and dynamics of sugarcane production

Sugarcane originates from tropical South- and Southeast Asia. Crystallized sugar, extracted from the sucrose stored in the stems of sugarcane, was known 5000 years ago in India. In the 7th century, the knowledge of growing sugarcane and producing sugar was transferred to China. Around the 8th century sugarcane was introduced by the Arabs to Mesopotamia, Egypt, North Africa and Spain, from where it was introduced to Central and South America. Christopher Columbus brought sugarcane to the Caribbean islands, today's Haiti and Dominican Republic. Driven by the interests of major European colonial powers, sugarcane production had a great influence on many tropical islands and colonies in the Caribbean, South America, and the Pacific. In the 20th century, Cuba played a special role as main supplier of sugar to the countries of the Former USSR. In the last 30 years, Brazil wrote a new chapter in the history of sugarcane production, the first time not driven by colonial powers and the consumption of sugar, but substantially driven by domestic policies fostering bioethanol production to increase energy self-reliance and to reduce the import bill for petroleum.

1.1. Regional distribution and dynamics of sugarcane production

World crop and livestock statistics collected and published by the Food and Agriculture Organization (FAO) of the United Nation are available for years since 1950. According to these data, world production of sugarcane at the mid of last century was about 260 million tons produced on around 6.3 million hectares, i.e. an average yield of just over 40 tons per hectare. Only 30 years later, in 1980, the global harvest of sugarcane had reached a level of some 770 million tons cultivated on about 13.6 million hectares of land with an average yield of 57 tons per hectare. Another nearly 30 years later, the estimates of sugarcane production for 2007 indicate more than doubling of outputs to 1525 million tons from some 21.9 million hectares harvested sugarcane. In summary, the global harvest of sugarcane had a nearly six-fold increase from 1950 to 2007 while harvested area increased 3.5 times. During the same period average global sugarcane yield increased from 41.4 tons per hectare in 1950 to 69.6 tons per hectare in 2007, i.e. a sustained average yield increase per annum of nearly 1%.

Figure 1 shows the time development and broad regional distribution of sugarcane production and area harvested.

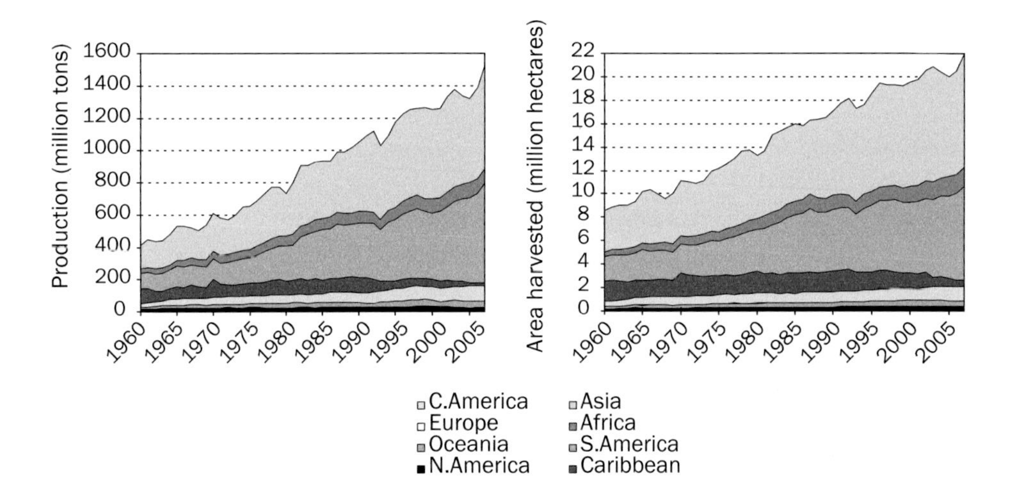

Figure 1. Global sugarcane production 1960-2007, by broad geographic region. a: production (million tons); b: area harvested (million hectares). Source: FAOSTAT, online database at http://www.fao.org, accessed July 2008.

Table 1 indicates the main global players in sugarcane production. The countries shown are listed in decreasing order of their sugarcane production in 2007. The table includes all those countries, which ranked at least once among the 10 largest global producers in past decades since 1950, and shows their global production rank for each period.

Table 2 indicates for the same countries level of production for respectively 1950 (three-year average for 1949-1951), 1960, etc., to 2000 (three-year average for 1999-2001), and for 2007. Table 3 presents associated harvested sugarcane areas.

In 1950, and still in 1960, India and Cuba were the two largest sugarcane producers in the world. India continued to dominate sugarcane production until 1980, when Brazil took over the first rank both in terms of area harvested and sugarcane output. Cuba maintained rank three among global sugarcane producers until 1991. Then, however, with the collapse of the USSR, Cuba's guaranteed sugar export market, the sugar industry in Cuba collapsed rapidly as well. As a result, sugarcane production in 2007 was only about one-eighth of the peak reached in 1990. Another example for the decline of Caribbean sugarcane industry is Puerto Rico, the world's seventh largest producer in 1950, where sugarcane cultivation became uneconomical and was completely abandoned in recent years.

Though the FAO lists more than 100 countries where sugarcane is cultivated, Table 2 and 3 indicate that global sugarcane production is fairly concentrated in only a few countries. The 15 top countries listed in Table 1 account for about 85 percent of the harvested sugarcane area in 2007, and for a similar level in 1950 and the other periods shown. The first three

Table 1. Rank of major producers of sugarcane, 1950-2007.

	2007	**1999-01**	**1989-91**	**1979-81**	**1969-71**	**1959-61**	**1949-51**
Brazil[1]	1	1	1	1	2	3	3
India[3]	2	2	2	2	1	1	1
China[1]	3	3	4	5	8	6	8
Thailand[1]	4	4	6	12	20	27	43
Pakistan[1]	5	5	7	7	6	9	12
Mexico[3]	6	6	5	4	4	4	6
Colombia[3]	7	9	9	8	11	7	5
Australia[1]	8	7	12	10	9	12	11
United States[2]	9	10	10	9	7	5	4
Philippines[3]	10	11	11	6	5	8	10
Indonesia[1]	11	12	8	11	12	11	18
South Africa[3]	12	13	13	13	10	15	13
Argentina[2]	13	14	14	14	13	10	9
Cuba[2]	17	8	3	3	3	2	2
Puerto Rico[2]	>100	88	56	40	21	13	7

Source: FAOSTAT, online database at http://www.fao.org, accessed July 2008; FAO, 1987.

[1] Countries that have significantly improved their rank in global production during the last five decades.

[2] Countries that have lost global importance in sugarcane production.

[3] Countries that occupied a rank in 2007 similar to their position in the 1950s.

countries – Brazil, India and China – produced more than 60 percent of the global sugarcane harvest in 2007; Brazil alone contributed about one-third. Somewhat lower, but similar ratios hold for sugarcane area harvested in 2007: the top three countries accounted for 58 percent of land harvested, Brazil for about 30%, which indicates that these countries enjoy sugarcane yields above the world average.

The dominance of Brazil in global sugarcane production and expansion – Brazil accounted for 75 percent of sugarcane area increases in the period 2000 to 2007 and two-thirds of global production increases in that period – derives from its experience and capability to respond to thriving international demand for transport fuels, which was recently triggered by measures to mitigate greenhouse gas emissions of the rapidly growing transport sector, concerns in developed countries to enhance energy security and lessen dependence on petroleum, and not the least the need of many developing countries to reduce import bills for fossil oil.

Table 2. Sugarcane production (million tons) of major producers, 1950-2007.

	2007	1999-01	1989-91	1979-81	1969-71	1959-61	1949-51
Brazil	514.1	335.8	258.6	147.8	78.5	56.6	32.2
India	322.9	297.0	223.2	144.9	128.7	87.3	52.0
China	105.7	75.1	63.9	33.8	19.6	15.0	8.0
Thailand	64.4	51.3	37.0	17.7	5.4	1.9	0.3
Pakistan	54.8	48.4	36.2	29.1	23.8	11.6	6.4
Mexico	50.7	46.1	40.8	34.4	33.3	18.8	9.8
Colombia	40.0	33.1	27.4	24.7	13.2	12.5	11.1
Australia	36.0	35.3	24.2	23.4	17.6	9.4	6.5
United States	27.8	32.1	26.6	24.5	21.4	16.0	13.5
Philippines	25.3	25.6	25.2	31.5	25.3	12.0	7.1
Indonesia	25.2	24.2	27.6	19.5	10.3	9.6	3.1
South Africa	20.5	22.1	18.9	17.3	14.6	8.2	4.7
Argentina	19.2	17.9	15.9	15.6	10.2	10.4	7.6
Cuba	11.1	34.2	80.8	69.3	60.5	58.3	44.5
Puerto Rico	0.0	0.1	0.9	2.0	5.0	9.4	9.7
Sum of above	1,317.5	1,078.2	907.1	635.5	467.1	337.0	216.5
World	1,524.4	1,259.4	1,053.5	768.1	576.3	413.0	260.8

Source: FAOSTAT, online database at http://www.fao.org, accessed July 2008; FAO, 1987.

Tables 1 to 3 point to two main factors that underlie the dynamics of sugarcane cultivation during the last four decades: a four-fold expansion of sugarcane acreage in South America between 1960 and 2007, and a collapse of sugarcane cultivation in the Caribbean sugar islands, especially important Cuba and Puerto Rico, which still held a substantial production share until the late 1980s. Solid growth of production and about three-fold expansion of sugarcane acreage since 1960 occurred in Asia mainly fuelled by rapid domestic demand increases for sugar in China and India. Fuel ethanol production from sugarcane has played a minor role in these dynamics with the exception of Brazil where it caused a large expansion.

An additional factor promoting the global expansion of sugarcane cultivation is the plant's efficient agronomic performance and its comparative advantage relative to sugar beets. While post-war self-reliance policies and protection of agriculture in developed countries supported an expansion of sugar beet cultivation areas until the late 1970s, the last three decades witnessed a gradual decline in harvested areas of sugar beet and increasingly a substitution of temperate sugar beets as a raw material for sugar production with tropical sugarcane (Figure 2). Regional changes of sugarcane cultivation are shown in Figure 3.

Table 3. Sugarcane area harvested (million hectares) in major producing countries, 1950-2007.

	2007	**1999-01**	**1989-91**	**1979-81**	**1969-71**	**1959-61**	**1949-51**
Brazil	6,712	4,901	4,092	3,130	1,830	1,400	1,307
India	4,830	4,197	3,699	3,073	2,486	2,428	2,011
China	1,225	1,171	1,230	722	566	279	414
Thailand	1,010	903	897	549	159	62	53
Pakistan	1,029	1,042	888	894	574	407	418
Mexico	680	628	556	520	483	352	325
Colombia	450	400	344	270	260	294	280
Australia	420	412	333	314	234	159	131
United States	358	412	374	306	282	184	176
Philippines	400	365	367	409	446	240	205
Indonesia	350	381	392	234	77	75	62
South Africa	420	392	272	252	181	96	110
Argentina	290	282	258	314	242	218	264
Cuba	400	1,015	1,372	1,246	1,254	1,218	1,097
Puerto Rico	0	3	16	25	61	129	133
Sum of above	18,574	16,504	15,089	12,257	9,134	7,539	6,986
World	21,896	19,476	17,729	14,708	11,025	8,946	8,302

Source: FAOSTAT, online database at http://www.fao.org, accessed July 2008; FAO, 1987.

1.2. Global significance of ethanol production from sugarcane

As shown in the previous analysis, for most of the 20th century sugarcane production took place in response to global demand for sugar, was largely conditioned by the heritage of colonial structures, and was greatly influenced by policy and trade agreements. With the launching of the PROALCOOL program in Brazil in the mid 1970s another important demand factor entered the scene, initially of national importance only. As a consequence of the program however Brazil became the largest sugarcane producer in the world and by now the largest exporter of transport bioethanol.

Figure 4 shows the dynamics of area expansion for sugarcane cultivation in Brazil and indicates the significant amount of land dedicated to ethanol production and the important role of the ethanol program in this process. The figure illustrates three phases that characterize the last three decades. In the first decade after launching the PROALCOOL program, i.e. during 1975 to 1986, there was a sharp increase in Brazilian sugarcane area, which is entirely due to the domestic feedstock demand of the ethanol program. Then, during 1986 to 2000, the figure suggests a growth of sugar production but a phase of stagnation in ethanol

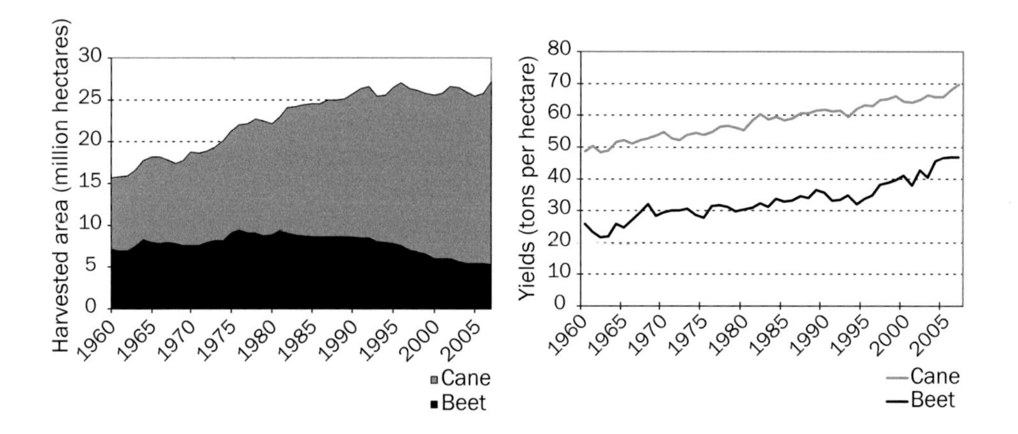

Figure 2. Harvested area and yields of sugarcane and sugar beet, 1960-2007. Source: FAOSTAT, online database at http://www.fao.org, accessed July 2008.

production, which has been attributed to various national and international factors, not the least a low price of petroleum. Finally, the most rapid expansion of sugarcane harvested areas occurred after 2000 and in particular during 2005 to 2008. This time ethanol demand to substitute for gasoline consumption became a driving force at the global level, with many countries seeking ways to cut greenhouse gas emissions and reducing dependence of their economies on imported fossil oil.

In recent years, biofuels have re-emerged as a possible option in response to climate change, and also to concerns over energy security. At the same time, many concerns among experts worldwide have been raised about the effectiveness to achieve these goals and the possible negative impacts on the poor, in particular regarding food security (Scharlemann and Laurance, 2008) and environmental consequences.

Recent sharp increases of agricultural prices have partly been blamed on rapid growth of biofuel production, especially maize-based ethanol production in the United States, which in 2007 absorbed more than a quarter of the US maize harvest. How important is sugarcane in this respect, and what fraction of the global sugar harvest is currently used for ethanol production?

Figure 5 shows world fuel ethanol production, which is dominated by two producers, the USA and Brazil. In 2008 these two countries contribute nearly 90 percent of total fuel ethanol production. Though detailed data on used feedstocks are difficult to obtain, it can be concluded that 45-50% of the world fuel ethanol production is based on sugarcane, requiring some 280 to 300 million tons of sugarcane from an estimated 3.75 million hectares harvested area (Table 4).

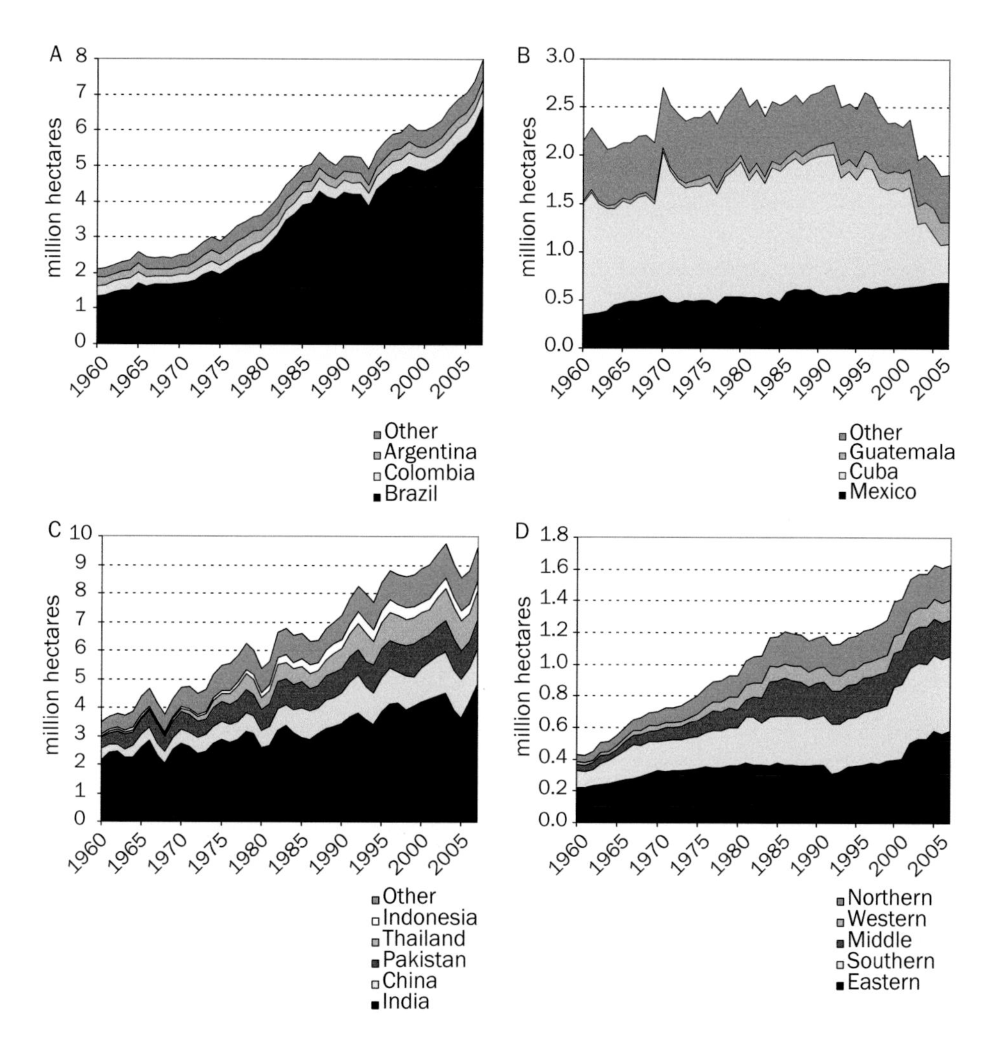

Figure 3. Change in sugarcane cultivation 1960-2007, by broad geographic region. a: South America (million hectares); b: Central America & Caribbean; c: Asia (million hectares); d: Africa (million hectares). Source: FAOSTAT, online database at http://www.fao.org, accessed July 2008.

Table 4 and 5 summarize the available data for two time points, 1969-71 and 2007. Apart from basic sugarcane statistics, the regional land-use significance of sugarcane is shown in terms of percentage of cultivated land used for sugarcane cultivation. For 1970, the region of Central America & Caribbean had the highest share where an estimated 7 percent of cultivated land was used for growing sugarcane. At that time, Brazil devoted 4.4 percent of cultivated land to sugarcane. In comparison, in year 2007 just over 10 percent of cultivated land were in use in Brazil to serve the sugar and ethanol industries. As a consequence, at the regional scale South America shows the highest share in 2007, now allocating 6.6 percent

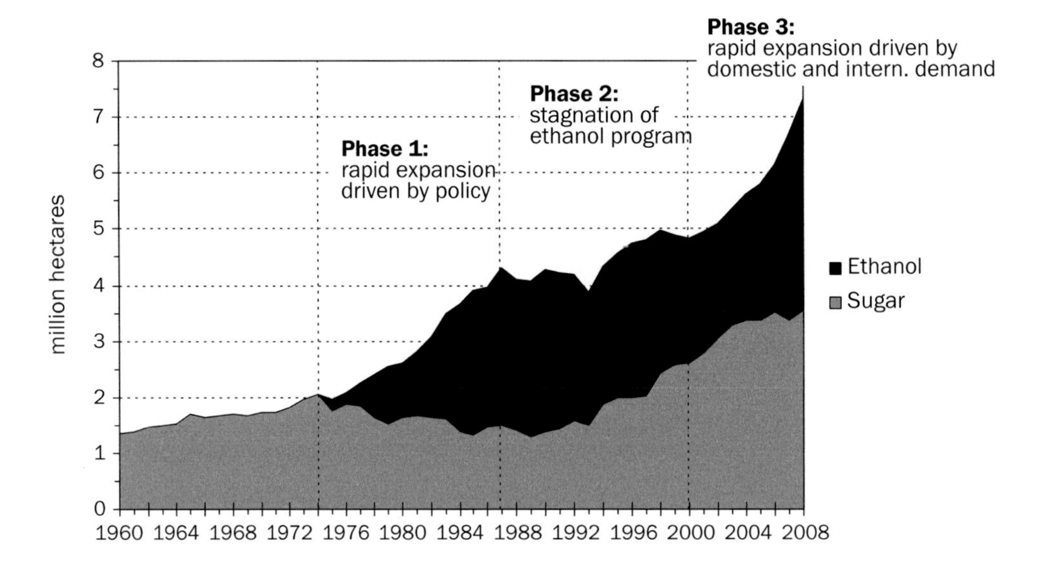

Figure 4. Use of Brazilian sugarcane land for ethanol and sugar production. Source: FAOSTAT, 2008; Conab, 2008a; Licht, 2007, 2008; calculation by authors.

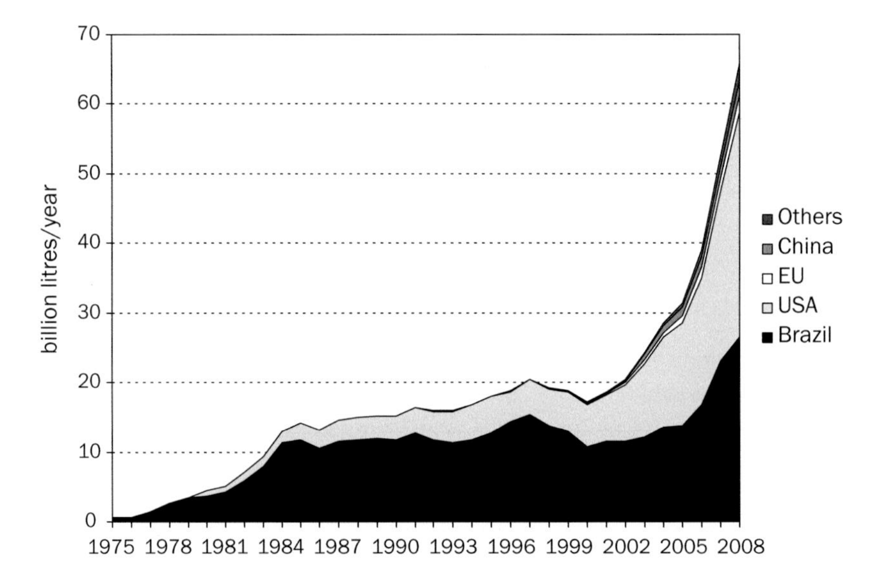

Figure 5. World fuel ethanol production (billion liters/year). Source: Licht, 2007 and 2008.

of total cultivated land to sugarcane. In comparison, the countries holding rank two and three in global production, India and China, devoted respectively 2.8 and 1.0 percent of cultivated land to sugarcane. The estimate for the global level amounts to 1.4 percent, i.e.

Table 4. Global significance of sugarcane production in 2007.

	Sugarcane			Cultivated land [1]	Sugarcane % of total cultivated	Sugarcane ethanol land	Ethanol % of sugarcane
	Harvested million ha	Production million tons	Yield tons/ha	million ha	percent	million ha	percent
North America	0.4	28	77.6	229.3	0.2	0	0
Europe & Russia	< 0.1	< 1	61.4	296.4	0.0	0	0
Oceania & Polynesia	0.5	40	79.9	54.8	0.9	0	0
Asia	9.6	639	66.4	577.1	1.7	< 0.1	< 1
Africa	1.6	92	56.8	239.3	0.7	< 0.1	< 1
Centr. Am. & Carib.	1.8	114	63.4	42.9	4.2	< 0.1	1
South America	8.0	611	76.5	121.9	6.6	3.6	45
Developed	0.9	67	78.9	580.4	0.1	0	0
Developing	21.0	1457	69.2	981.3	2.1	3.8	17.8
World	21.9	1524	69.6	1561.7	1.4	3.8	17.1
Brazil	6.7	514	76.6	66.6	10.1	3.5	50
India	4.8	323	72.6	169.7	2.8	< 0.1	n.a.
China	1.4	106	86.2	140.0	1.0	< 0.2	n.a.
Thailand	1.0	64	63.7	17.8	5.7	< 0.1	3
Pakistan	1.0	55	53.2	22.1	4.7	0	n.a.

Source: FAOSTAT, 2008; Licht, 2007, 2008; calculation by the authors.
[1] Estimates of cultivated land refer to year 2005.

Table 5. Global significance of sugarcane production in 1969-71.

	Sugarcane			Cultivated land	Sugarcane % of total cultivated
	Harvested million ha	Production million tons	Yield tons/ha	million ha	percent
North America	0.2	21	89.8	243.4	0.1
Europe & Russia	< 0.1	< 1	72.1	378.3	0.0
Oceania & Polynesia	0.3	20	75.0	46.2	0.6
Asia	4.6	227	49.5	448.7	1.0
Africa	0.7	47	66.2	180.5	0.4
Centr. Am. & Carib.	2.5	132	53.9	34.9	7.0
South America	2.5	128	51.7	90.6	2.7
Developed	0.5	42	82.8	667.9	0.1
Developing	10.2	534	52.2	754.6	1.4
World	10.7	576	53.7	1422.6	0.8
Brazil	1.8	78	45.9	41.3	4.4
India	2.5	129	48.9	164.7	1.5
China	0.6	20	41.3	102.5	0.6
Thailand	0.1	5	44.5	13.7	0.4
Pakistan	0.6	24	39.9	19.3	3.0

Source: FAOSTAT, 2008.

sugarcane harvested was 22 million hectares out of 1562 million total cultivated land. In comparison, the share of sugarcane in global cultivated land was 0.8 percent in 1970, which means that nearly a doubling of the global significance of sugarcane has occurred in the last three decades.

At first glance, the rather low percentage of global cultivated land occupied by sugarcane suggests that sugarcane area expansion and associated land competition has had little influence on food supply. Yet, this may be misleading for two reasons: (1) sugarcane is cultivated either under irrigation (e.g. India and Pakistan) or in rain-fed tropical areas with ample rainfall. Hence land productivity in areas suitable for rain-fed sugarcane production is typically much higher than for cultivated land in cooler climates or arid sub-tropical and tropical agriculture; and (2) large parts of the world cannot grow sugarcane for climatic reasons and the impact in climatically suitable areas is therefore more significant, as shown in Table 6.

Table 6. Global significance of sugarcane production in 2007 revisited.

	Sugarcane harvested area	Cultivated land		Sugarcane harvested	
		Total	With sugarcane potential	% of total cultivated land	% of cultivated land with sugarcane potential
	million ha	million ha	million ha	percent	percent
North America	0.4	229.3	17.6	0.2	2.0
Europe & Russia	< 0.1	296.4	0.8	0.0	0.1
Oceania & Polynesia	0.5	54.8	2.5	0.9	19.5
Asia	9.6	577.1	213.3	1.7	4.5
Africa	1.6	239.3	81.6	0.7	2.0
Centr. Am. & Carib.	1.8	42.9	28.0	4.2	6.4
South America	8.0	121.9	90.2	6.6	8.9
Developed	0.9	580.4	19.5	0.1	4.4
Developing	21.0	981.3	414.4	2.1	5.1
World	21.9	1561.7	434.0	1.4	5.0
Brazil	6.7	66.6	57.3	10.1	11.7
India	4.8	169.7	70.1	2.8	6.8
China	1.4	140.0	12.4	1.0	11.3
Thailand	1.0	17.8	17.0	5.7	5.9
Pakistan	1.0	22.1	15.6	4.7	6.4

Source: FAOSTAT, 2008; Fisher *et al.*, 2008.

The global analysis clearly shows that the most significant and relevant land use change dynamics related to sugarcane in the last decades have taken place in Brazil. In the following we take a short look at the Brazilian development and some issues and questions this development has raised.

1.3. Sugarcane and land use change dynamics in Brazil

Brazil has the largest area under sugarcane cultivation in the world, being responsible for approximately one third of the global harvested area and production. For the year 2007, 6.7 million hectares were harvested with a production of 514 million tons of sugarcane

(FAOSTAT, 2008). The land use change into sugarcane production is part of the history of the country, dating short after Portuguese colonization during the 16th century. Since then, the crop has maintained its characteristic of a monoculture with high elasticity of supply, expanding rapidly in response to market stimuli (Tercil *et al.*, 2007). The first establishment phase of the crop over native vegetation aimed to provide sugar to the growing European market during colonial times, during this period plantations were established in the North-East and South-East of the country where agro-ecological conditions are highly favorable for the growth of tropical grasses such as sugarcane (e.g. see Figure 2.10 in next section).

From 2000 to 2007, an impressive pace of approximately 300 thousand hectares of land was converted into sugarcane every year (FAOSTAT, 2008). This already phenomenal rate of conversion is being surpassed by recent projections for the 2007/08 harvest season, which indicate an expansion of 650 thousand hectares in Brazil (Conab, 2008a). Most of the recent expansion in sugarcane area has occurred in São Paulo state (Conab, 2008a). From 1995 to 2007, there was a 70% enlargement of the sugarcane area in São Paulo, from 2.26 million ha to 3.90 million ha, which represents 58% of the Brazilian area under sugarcane (IEA, 2007). In response to a greater demand for ethanol, São Paulo is also the region where most of the land use change into sugarcane plantation is expected to take place in the near future (Goldemberg *et al.*, 2008). The projected expansion of sugarcane for the 2007/08 harvest season is 350 thousand hectares, i.e. 54% of the Brazilian total (Conab, 2008b). Therefore, we further discuss the aspects of land use change in Brazil with special attention on São Paulo as an example of intensive conversion of other land uses into sugarcane monocultures.

The basis for the success of the crop in the South-East of Brazil is the favorable environmental conditions in terms of temperature, radiation, precipitation, soil characteristics and relief that match the crop physiological requirements. The potential to achieve high yields, today an average near 80 t/ha (Conab 2008b), has diluted fixed production costs and has established Brazilian ethanol as one of the most competitive bio-fuel options with an estimated cost of US$ 0.21/liter (Goldemberg, 2007).

1.4. What are the drivers for these changes in Brazil?

The main drivers for the recent expansion of sugarcane in Brazil, particularly São Paulo, were market opportunities created by the international demand for sugar and ethanol in conjunction with national policies that promoted ethanol production and commercialization. During these periods, intense and initially heavily subsidized investments (e.g. PROALCOOL in mid 70's) allowed the development of a solid industrial capacity and know-how (Goldemberg, 2006). The historical background of sugarcane as a traditional land use and the investments in the ethanol production chain created ideal conditions for the development of indigenous technologies on agronomical (e.g. plant nutrition, management and high yielding genetic material) and industrial aspects of production. For example, the flexibility to shift between sugar and ethanol production (mixed production units) mitigates fluctuations on the demand side, which makes the business highly attractive as a land use option. Currently, mixed production units process 85.4% of Brazil's industrialized sugarcane (Conab, 2008b). Another aspect that favors rapid expansion of sugarcane in Brazil is the current land tenure structure in this agri-business. There is a large concentration of land in the hands of the industry, 67% of Brazilian sugarcane producing areas (Conab, 2008b). The operation of extensive sugarcane farms reduces the cost of production through economy of scale (Goldemberg, 2006) contributing to the overall competitiveness of sugarcane production in relation to other land uses options. Finally, the environmental conditions in vast areas of Brazil's arable land are adequate not only for achieving high sugarcane yields (see Figure 10) but also high sucrose concentrations, i.e. a cool and dry winter period in São Paulo favors accumulation of sugar, which increases industrial efficiency (Conab, 2008b). In combination, these favorable biophysical conditions and socio-economical historical aspects produced a setting for effective response to political and market stimuli explaining the rapid expansion of sugarcane monoculture in Brazil.

1.5. What have been the impacts on environmental parameters?

The recent boom of ethanol production has drawn international attention to the environmental impacts of land conversion into sugarcane monocultures. Site-specific biophysical and socio-economical aspects largely determine the impacts of land use change. The conversion of land use, its susceptibility to land degradation and the choice of agronomic and agro-processing technologies for sugarcane production and conversion determine the magnitude of impacts on environmental quality at the local level. Major areas of concern include deforestation and threats to biodiversity, environmental pollution and competition with food crops.

1.5.1. Deforestation and threats to biodiversity

The expansion of sugarcane could increase deforestation rates either 'directly' by intruding in areas of native non-protected forest areas or 'indirectly' by forcing other land uses (e.g. displaced livestock production and agricultural crops such as soybeans) to open up new land. Past surges of sugarcane expansion in Brazil are not regarded as a major cause of deforestation (Martinelli and Filoso, 2008). The current sugarcane area represents only 2.5% of the 264 million ha of agricultural land use in Brazil, of which nearly 200 million ha are pastoral lands. The hotspots of deforestation in the Amazon region, however have a low suitability for sugarcane production and are not directly threatened by the current sugarcane expansion (Smeets *et al.*, 2008). Amazon deforestation has been caused mainly by conversion to pastoral lands for livestock production and, more recently, also for expansion of soybean production (Fearnside, 2005).

From 1988 to 2007 the average rate of expansion of sugarcane was 0.14 million ha/year when rates of Amazon deforestation ranged from ~1.1 to 2.9 million ha per year (Fearnside, 2005) indicating that sugarcane expansion is by far insufficient to have forced 'direct' or the 'indirect' reallocation of pasture and soybeans northwards intruding into Amazon rainforests.

Currently, the savannah region ('Cerrados'), considered a world bio-diversity hotspot (Myers *et al.*, 2000), is the ecosystem most threatened by sugarcane expansion in Brazil as it is situated on the frontier of agricultural expansion and has at least partly excellent cultivation potentials (Klink and Machado 2005; Smeets *et al.* 2008). The Cerrado is characterized by high biodiversity (e.g. >6.5 thousand plants species from which 44% are endemic to the biome) and has suffered rates of conversion to either cultivated pasture land or to crop cultivation land that are higher than the deforestation rates in Amazon (Conservation International, 2008; Klink and Machado, 2005). In 2002, nearly 40% of a total of about 205 million ha of Cerrado had already been converted (Table 7), mainly into pastures and cash-crops such as soybeans (Machado *et al.*, 2004; Sano *et al.*, 2008).

From the early 1970s to 2000 around 0.36 million ha of Cerrado vegetation were lost in São Paulo (Florestar, 2005). However, from 2001 to 2005, total native vegetation areas in this state were maintained at about 3.15 million ha suggesting that more recent sugarcane expansion was not a major lever of deforestation during this period. Nevertheless, specific ecological systems such as riparian forests were highly affected in regions of intensive sugarcane production to give way to cropping areas (Martinelli and Filoso, 2008). In major watersheds in São Paulo State, where pastures and sugarcane are the main land uses, it is

Table 7. Land use shares of the Brazilian Cerrado region in 2002 (Adapted from Sano *et al.*, 2008 and Ministério do Meio Ambiente, 2007).

Land use classes	**Area (million ha)**	**Percent of total**
Native areas	124	60%
Native forest	75	37%
Native non-forest [1]	48	24%
Anthropic areas	80	39%
Cultivated pastures	54	26%
Agriculture	21	10%
Reforestation	3	2%
Urbanized plus mining	1	<1%
Water	1	1%
Total cerrado area	205	100%

[1] The 48 million ha of non-forested areas are estimated to include 28 million ha of native pastures (Ministério do Meio Ambiente, 2007).

shown that 75% of the riparian vegetation (a reservoir of biodiversity and a buffer against sedimentation of water bodies) had disappeared (Silva *et al.*, 2007).

1.5.2. Air, water and soil pollution and degradation

During the past surges of sugarcane expansion, cases of environmental pollution were identified at different stages of production and industrialization. The impacts on air, water and soil quality largely depend on the choices of technologies applied in agronomic and agro-processing practices. Beyond carbon releases and biodiversity losses caused by land conversion (discussed above), the main environmental effects concern air pollution from pre-harvest sugarcane burning, water pollution from cultivation and processing of sugarcane, and soil erosion and compaction as a consequence of sugarcane cultivation.

For example, air quality is highly compromised by the common practice of sugarcane burning, a technique used before harvest to facilitate manual cutting. The emission of pollutants during the dry months of the year, when harvest occurs in São Paulo, has direct negative impacts on health (e.g. respiratory disorders mainly in children and elderly citizens). It promotes erosion of topsoil, causes loss of nutrients and leads to soil compaction (Tominaga *et al.*, 2002; Cançado *et al.*, 2006; Ribeiro, 2008).

Soil degradation through erosion and compaction are also considered a problem in sugarcane fields, which are under intense mechanization during soil cultivation and harvesting (Martinelli and Filoso, 2008). Soil compaction is a consequence of the traffic of heavy machinery in conjunction with the lack of implementation of best management cultivation practices (Naseri *et al.*, 2007). Compaction exacerbates erosion problems because soil porosity is reduced, which decreases water infiltration and increases runoff (Oliveira *et al.*, 1995; Martinelli and Filoso 2008). The main periods when soil remains bare and subjected to erosive forces by rain and winds are (1) during the process of land conversion, (2) between crop harvesting and subsequent canopy closure, and (3) during re-planting of sugarcane fields every 5-6 years. The conversion of natural vegetation and extensive pastures (which are less intensively managed) into sugarcane increases the risk soil degradation (Politano and Pissarra, 2005). Erosion rates of 30 Mg of soil/ha.year were estimated for sugarcane fields in the São Paulo State in comparison with less than 2 Mg/ha.year for pastures and other natural vegetation (Sparovek and Schnug, 2001). Soil erosion in poorly managed sugarcane areas also causes sediment deposition into water reservoirs, wetlands, streams and rivers (Politano and Pissarra, 2005). This is aggravated by the transport of fertilizer and agro-chemical residues that directly compromise water quality (Corbi *et al.*, 2006).

Water pollution has been a severe environmental problem in sugarcane production regions until early 80's in Brazil when legislation was implemented to ban direct discharge of vinasse (Martinelli and Filoso, 2008; Smeets *et al.*, 2008). The main industrial sources of pollutants of sugarcane industry are wastewater from washing of stems before processing and vinasse produced during distillation. These by-products have a large potential of water contamination due to a high concentration of organic matter, which increases the biochemical oxygen demand (BOD_5) of water bodies receiving such effluents (Gunkel *et al.*, 2007). While the Brazilian standards for wastewater emission are BOD_5 of 60 mg/l, values for wastewater from cane washing are up to 500 mg/l and > 1.000 mg/l for vinasse (Gunkel *et al.*, 2007; Smeets *et al.*, 2008). In addition, agro-chemicals residues have been found as a important component of water pollution in areas of intense sugarcane production (Corbi *et al.*, 2006; Silva *et al.*, 2008).

1.5.3. Land use and competition with food crops

A major area of concern is the threat to food security (Goldemberg *et al.*, 2008). Rapid expansion of sugarcane areas could potentially reduce the availability of arable land for the cultivation of food and feed crops causing a reduction in their supply and increase of food prices. Fast rates of expansion of sugarcane in São Paulo state in the mid 70s at the expense of maize and rice cropping areas seem to have had a short-term impact on regional food supply and prices (Saint, 1982). However, the recent sugarcane expansion in São Paulo from mid 90's has not compromised food crop production as most of the expansion intruded in pastoral lands (Figure 6).

For Brazil as a whole, in the 2006/07 season, nearly two thirds of sugarcane expansion occurred at the expense of pastures (0.42 million ha) in comparison with one quarter coming from land under crop cultivation (Conab, 2008b). This conversion of pastures into sugarcane

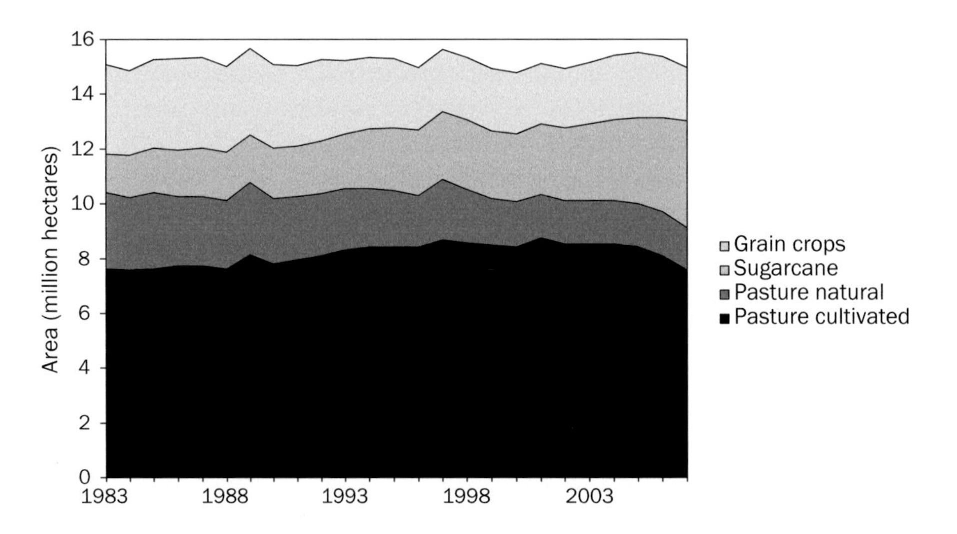

Figure 6. Evolution of areas of sugarcane, pasture and grain crops in São Paulo State. Source: IEA, 2007; Conab, 2008c. Note: The total area of São Paulo State is 24.8 million ha.

areas is explained by their relative abundance (200 million ha) as well as occurrence adjacent to existing sugarcane estates (Goldemberg *et al.*, 2008).

The area of main grain crops has decreased by 0.9 million ha in the State of São Paulo from early 80's to 2005 (Conab, 2008c), while sugarcane area expanded nearly 1.7 million ha (IEA 2007), Figure 7. At the national level the magnitude of these regional land use changes is diluted (Figure 8) as the total area of major crops, including sugarcane, is about 50 million ha (Conab, 2008c). By far more important than sugarcane has been the rapid expansion of soybeans in Brazil, from less than 10 million hectares in the early 1980s to around 23 million hectares, more than a third of all cropping land.

1.6. Lessons from Brazilian sugarcane land development dynamics

The learning experience with deploying sugarcane based ethanol production in Brazil during the last 30 years has put the country in a unique position to respond to the current wave of energy systems developments, particularly renewable transport fuels. As to land use, the following conclusions can be summarized:

- There was a very rapid and large land use change into sugarcane production in Brazil in the last 30 years, particularly in São Paulo State.
- Main drivers for the expansion of sugarcane areas were a combination of favorable biophysical conditions, a historical foundation of logistical and technological conditions to respond to market opportunities, national policies giving incentives to the sugarcane

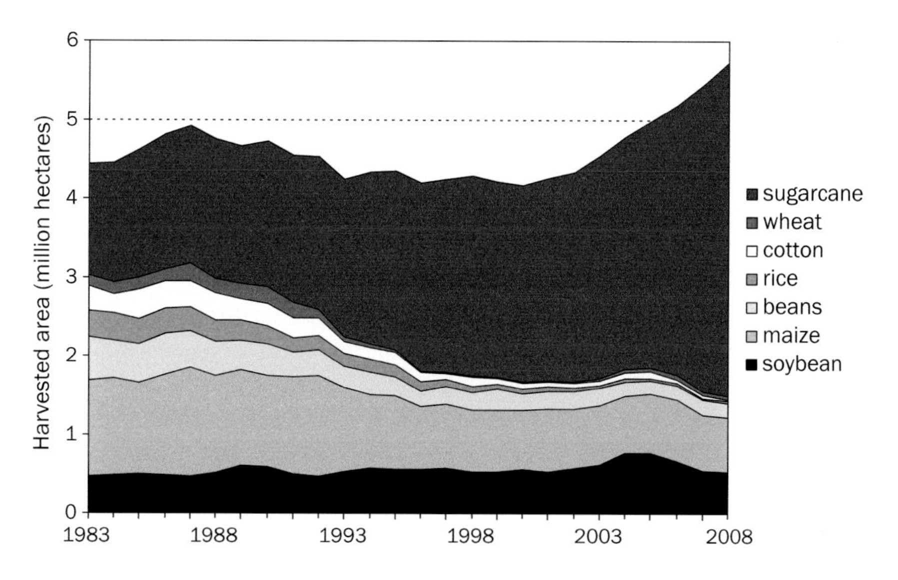

Figure 7. Area of selected crops in São Paulo. Source: Conab, 2008c.

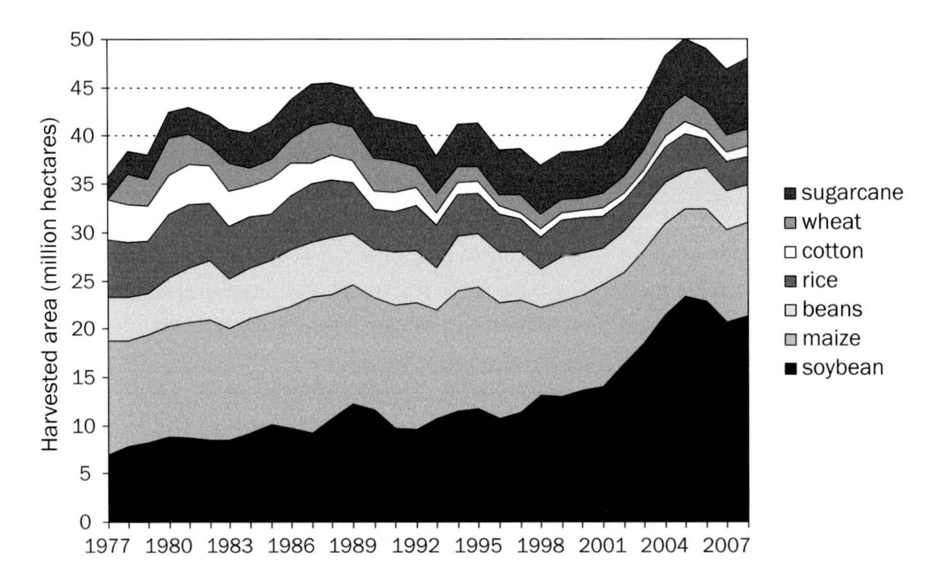

Figure 8. Area of selected crops in Brazil. Source: Conab, 2008c.

agri-business, and a growing demand for sugar and bioethanol, setting favorable conditions to benefit from economies of scale.

- The trend in sugarcane area expansion continues at record rates, now fostered by both the domestic and international demand for ethanol.
- The savannah ecosystem (Brazilian 'Cerrados') is the current frontier of sugarcane expansion.
- There are risks of environmental degradation in different stages of sugarcane production and processing. Negative impacts have been caused by the lack of implementation of best management practices and ineffective legislation and control. Examples from São Paulo state indicate that environmental sustainability of sugarcane production and processing has been substantially improved during the last three decades. Nevertheless, further improvements are necessary.
- While more effective and environmentally less harmful technologies are now available, there is nevertheless a risk of affecting biodiverse ecosystems of the savannah region. Strict regulation and enforcement are needed to safeguard against environmental losses, for example by guaranteeing the protection and recuperation of specific biomes such as the Cerrado and riparian forests.

2. Global potential for expansion of sugarcane production

2.1. Future land requirements for food and feed

Several inter-linked processes determine the dynamics of world food demand and supply. Agro-climatic conditions, availability of land resources and their management are clearly key aspects, but they are critically influenced by regional and global socio-economic pressures including current and projected trends in population growth, availability and access to technology, market demands and overall economic development.

While climate and farm management are key determinants of local food production, agro-economics and world trade combine to significantly shape regional and global agricultural land use. Catering to consumers and industries in OECD countries is an important driver for agricultural activities in well-resourced developing countries. Computations of current and future cultivated land were carried out by assessing land potential with the global Agro-ecological Zones model (GAEZ) and economic utilization with IIASA's world food system model (Fischer *et al.*, 2002; Fischer *et al.*, 2005). In 2000 about 1.5 billion hectares of arable land were in use for food, fiber and fodder crop production, or roughly 10% of all available land on earth. Of these, about 900 million hectares were in developing countries. By 2050, under a IIASA designed plausible global socio-economic development scenario (Grübler *et al.*, 2006; Tubiello and Fischer, 2006; Fischer *et al.*, 2006), for developed countries a slightly lower level of cultivated land use was projected compared to 2000, i.e. a modest net decrease in land under cultivation for food and feed crops was projected, while additional production resulted from increased productivity and input use. In developing countries, by contrast, cultivated land in 2050 was projected to increase by roughly 190 million ha (+21%) relative to year 2000. In the scenario, most of this additional cropland is brought into use in Africa (+85 million ha, or +42%) and Latin America (70 million ha, or +41%).

From a range of alternative scenario runs predicting world food system development (Fischer *et al.*, 2002; 2005) it can be concluded that global food and feed demand will require some additional land to be used for cultivation, depending on socioeconomic scenario in the range of 120-180 million hectares, notably in developing countries. Therefore, when adopting a 'food first' paradigm, to realize a substantial contribution of agricultural biomass to energy sources would necessitate (1) focused efforts of national and international R&D institutions and extension services to enable sustainable agricultural production increases on current agricultural land, which go beyond 'business as usual' trends and expectations, in particular to mobilize undeveloped agricultural potentials on the African continent, and (2) tapping into resources currently not or only extensively used for cultivation or livestock production, e.g. certain grass, scrub and woodland areas where environmental and social impacts might be regarded as acceptable. For this reason, we next look into the question as to how much land, where and under what current uses, could be potentially available for expanding global sugarcane production.

2.2. AEZ assessment of land suitable for sugarcane production

2.2.1. AEZ background

The range of uses that can be made of land for human needs is limited by environmental factors including climate, topography and soil characteristics, and is to a large extent determined by demographic and socioeconomic drivers, cultural practices, and political factors, e.g. such as land tenure, markets, institutions, and agricultural policies.

The Food and Agriculture Organization of the United Nations (FAO) with the collaboration of IIASA, has developed a system that enables rational land-use planning on the basis of an inventory of land resources and evaluation of biophysical limitations and production potentials of land. This is referred to as the Agro-ecological Zones (AEZ) methodology.

The AEZ methodology follows an environmental approach; it provides a standardized framework for the characterization of climate, soil and terrain conditions relevant to agricultural production. Crop modeling and environmental matching procedures are used to identify crop-specific limitations of prevailing climate, soil and terrain resources, under assumed levels of inputs and management conditions. This part of the AEZ methodology provides maximum potential and agronomically attainable crop and biomass yields globally at 5-minute latitude/longitude resolution grid-cells.

2.2.2. Land suitability for sugarcane

Sugarcane belongs to the crops with C4 photosynthetic pathway; it is adapted to operate best under conditions of relatively high temperatures and, in comparison to C3 pathway crops, has high rates of CO_2 exchange and photosynthesis, in particular at higher light intensities.

Sugarcane is a perennial with determinate growth habit; its yield is located in the stem as sucrose and the yield formation period is about two-thirds to three quarters of its cultivated life span. Climatic adaptability attributes of sugarcane qualify it as being most effective in tropical lowland and warm subtropical climates; it does particularly well in somewhat drier zones under irrigation, but is sensitive to frost. A short dry and moderately cool period at the end of its cultivation cycle significantly increases sugar content at harvest.

Ecological requirements of sugarcane include warm, sunny conditions and adequate soil moisture supply during most of its cultivation cycle. Sugarcane prefers deep, well drained, well structured and aerated loamy to clayey fertile soils. Ideal pH ranges are between 5.5 and 7.5.

2.2.3. AEZ procedures applied for sugarcane

Box 1 summarizes the AEZ methodology and information flow as applied for the assessment of global sugarcane potentials.

Box 1. AEZ procedures (see Figure 9).

Land Utilization Type (LUT): The AEZ procedures have been used to derive by grid-cell potential biomass and yield estimates for rain-fed sugarcane production under high level inputs/advanced management, which includes main socio-economic and agronomic/farm-management components:

The farming system is (1) market oriented; (2) commercial production of sugar and bioethanol are management objectives, and (3) production is based on currently available yielding cultivars, is fully mechanized with low labor intensity, and assumes adequate applications of nutrients and chemical pest, disease and weed control.

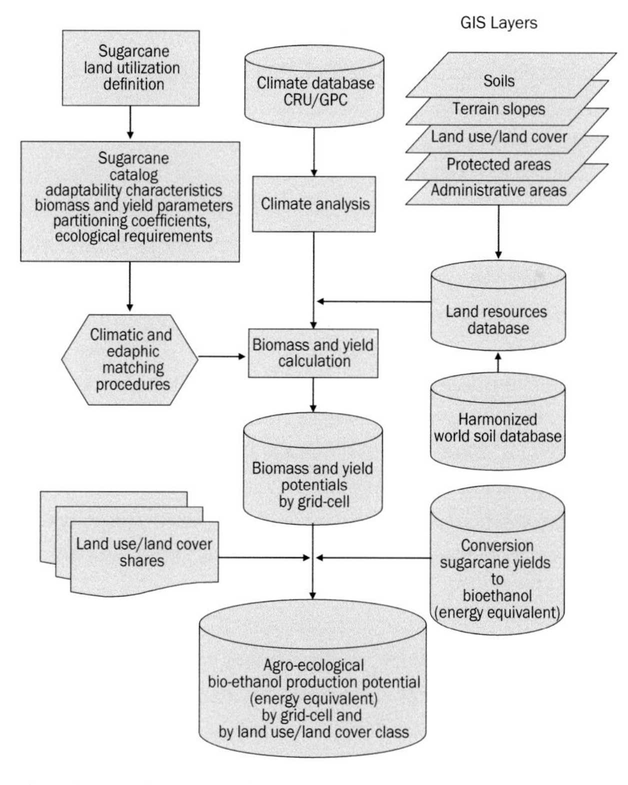

Figure 9. AEZ methodology: information flow and integration.

The quantified description of sugarcane LUTs include characteristics such as vegetation period, ratoon practices, photosynthetic pathway, photosynthesis in relation to temperature, maximum leaf area index, partitioning coefficients, and parameters describing ecological requirements of sugarcane produced under rain-fed conditions.

Climatic data: Climate data are from the Climate Research Unit (CRU CL 2.0 (New *et al.*, 2002, CRU TS 2.1; Mitchell and Jones, 2005), and precipitation data from VASClimO (Global Precipitation Climatology Centre - GPCC). Average climate and historical databases were used to quantify: (1) the length of growing period parameters, including year-to-year variability, and (2) to estimate for each grid-cell by crop/LUT, average and individual years agro-climatically attainable sugarcane yields.

Soils data: Spatial soil information and attributes data is used from the recently published Harmonized World Soil Database (FAO, IIASA, ISRIC, ISSCAS & JRC, 2008)

Terrain data: Global terrain slopes are estimated on the bases of elevation data available from the Shuttle Radar Topography Mission (SRTM) at 3 arc-second resolution

Land use/land cover: Potential yields, suitable areas and production were quantified for different major current land cover categories (Fischer *et al.*, 2008). The estimation procedures for estimating seven major land-use and land cover categories are as follows: Cultivated land shares in individual 5' grid cells were estimated with data from several land cover datasets: (1) the GLC2000 land cover regional and global classifications (http://www-gvm.jrc.it/glc2000), (2) the global land cover categorization, compiled by IFPRI (IFPRI, 2002), based on a reinterpretation of the Global Land Cover Characteristics Database (GLCC) ver. 2.0, EROS Data Centre (EDC, 2000) (3) the Forest Resources Assessment of FAO (FAO, 2001), and global 5' inventories of irrigated land (GMIA version 4.0; FAO/University of Frankfurt, 2006). Interpretations of these land cover data sets at 30-arc-sec. were used to quantify shares of seven main land use/land cover, consistent with land use estimates of published statistics. These shares are: cultivated land, subdivided into (1) rain-fed and (2) irrigated land, (3) forest, (4) pasture and other vegetation, (5) barren and very sparsely vegetated land, (6) water, and (7) urban land and land required for housing and infrastructure.

Protected areas: The principal data source of protected areas is the World Database of Protected Areas (WDPA) (http://www.unep-wcmc.org/wdpa/index.htm.) Two main categories of protected areas are distinguished: (1) protected areas where restricted agricultural use is permitted, and (2) strictly protected areas where agricultural use is not permitted.

Land resources database: Spatial data linked with attribute information from soils, terrain, land use and land cover, and protected areas are combined with an administrative boundary GIS layer in the land resources database

Climate analysis: Monthly reference evapotranspiration (ETo) has been calculated according to Penman-Monteith. A water-balance model provides estimations of actual evapotranspiration (ETa) and length of growing period (LGP). Temperature and elevation are used for the characterization of thermal conditions, e.g. thermal climates, temperature growing periods (LGP_t), and accumulated temperatures. Temperature requirements of sugarcane were matched with temperature profiles prevailing in individual grid-cells. For grid-cells with an optimum or sub-optimum match, calculations of biomass and yields were performed.

Edaphic modifiers: The edaphic suitability assessment is based on matching of soil and terrain requirements of the assumed sugarcane production systems with prevailing soil and terrain conditions.
Land productivity for rain-fed sugarcane: The combination of climatic and edaphic suitability classification provides by grid-cell potential biomass and yield estimates for assumed production conditions

2.3. Agro-ecological suitability of sugarcane – risks and opportunities of expansion

Figure 10 presents a map of climatically attainable relative yields for rain-fed conditions, normalized to a range of 0 (i.e. no yield possible) to 1 (i.e. geographical locations where highest rain-fed yields would be obtained). According to the AEZ assessment, the most suitable climates are found in the southeastern parts of South America, e.g. including São Paulo State in Brazil, but also large areas in Central Africa as well as some regions in Southeast Asia. Very wet areas with low temperature seasonality such as parts of the Amazon basin[1] produce substantially lower yields due to lower sugar content, high pest and disease incidence combined with lower efficacy of control, and in extreme wet areas difficulties with field operations and harvest. Note that in India and Pakistan, the world's second and fifth largest producers of sugarcane, irrigation is needed to exploit the thermal and radiation resources in these countries for sugarcane cultivation.

[1] Conditions in the equatorial parts of Africa differ substantially in wetness as compared to parts of the Amazon basin and provide from climate viewpoint better sugar yields.

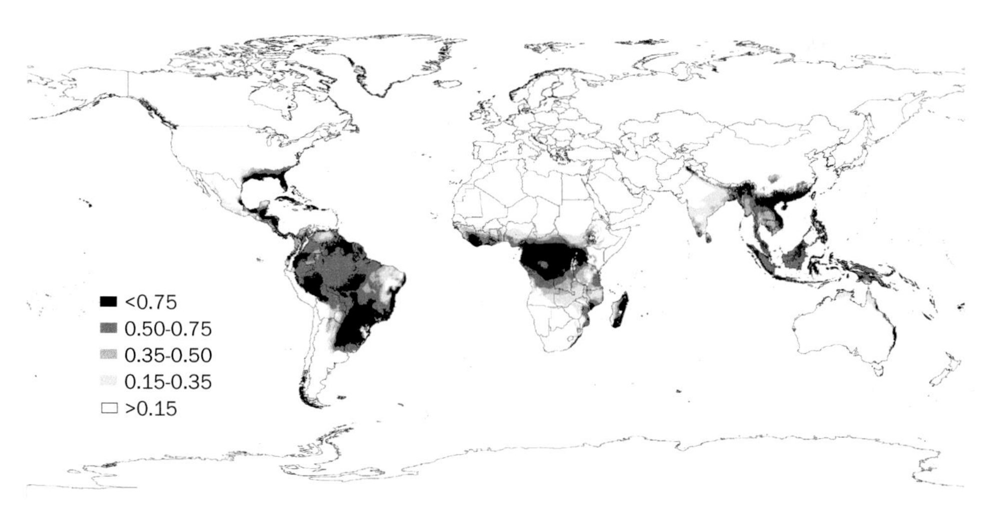

Figure 10. Normalized agro-climatically attainable yield of rain-fed sugarcane. Source: Fisher *et al.*, 2008, IIASA. Note: Maximum attainable yields in this global map are about 15 tons sugar per hectare.

Table 8 summarizes by region the current distribution of cultivated land, the land harvested for sugarcane in 2007, and the area of current cultivated land assessed as very suitable (VS), suitable (S) and moderately suitable (MS). Globally, the currently harvested 22 million hectares of land for sugarcane compare to the potential of 28 million hectares VS-land and 92 million hectares rain-fed S-land. In other words, of currently 1550 million hectares cultivated land about 120 million hectares is very suitable or suitable for rain-fed sugarcane cultivation, with the majority of this land located in developing countries of Africa (28 million hectares), Asia (34 million hectares) and South America (40 million hectares).

The Brazilian experience has shown that a major land source of sugarcane expansion was from pastures. The assessment of sugarcane suitability in current grass, scrub, wood land concluded that some 130 million hectares of this land would be very suitable or suitable for rain-fed sugarcane production, of which 48 million hectares were found in Sub-Saharan Africa and 69 million hectares in South America; Brazil accounts for nearly half this potential (Table 9). There is only very little potential of this kind, about 7 million hectares, in Asia as all the vast grasslands of Central Asia are too cold and too dry for rain-fed sugarcane production.

The maps for South America and Africa shown in Figure 11 indicate the suitability of climate, soil and terrain conditions for rain-fed sugarcane production. The respective suitability class is shown for areas where 50 percent or more of a grid-cell of 5' by 5' latitude/longitude is currently used as cultivated land and/or is covered by grass, scrub or woodland ecosystems. Hence, it shows the suitability of land where a substantial fraction is non-forest ecosystems. This geographical filter was used to indicate the distribution of land for potential sugarcane expansion, i.e. areas where further expansion of sugarcane would not cause direct deforestation and, provided the biodiverse native Cerrado ecosystem can be protected, would not create associated major risks for biodiversity and substantial carbon debts as is the case with forest conversion.

The maps shown in Figure 12 indicate the suitability of climate, soil and terrain conditions for rain-fed sugarcane production in areas where 50 percent or more of each grid-cell of 5' by 5' latitude/longitude is classified as forest or protected land, highlighting land at risk of undesirable conversion 'hot spots' due to its suitability for sugarcane expansion. Unlike the areas shown in Figure 11, conversion of these forest and protected areas would likely be associated with high environmental impacts.

While legally protected areas, both forests and non-forest ecosystems, are less exposed to conversion, unprotected forest areas with good suitability for rain-fed sugarcane cultivation are of particular concern due to possible severe environmental impacts. The AEZ methodology was therefore used to assess the magnitude and geographical distribution of unprotected forest areas. A summary of results by region is provided in Table 10.

Table 8. Suitability of current cultivated land for rain-fed sugarcane production.

	Cultivated land	Land potentially suitable, of which			Share of VS+S in cultivated land	Sugarcane harvested (2007)
		Very suitable (VS)	Suitable (S)	Moderately suitable		
	million ha	million ha	million ha	million ha	percent	million ha
North America	230	2.7	4.6	7.7	3.1	0.4
Europe & Russia	305	0	0	0	0.0	0.0
Oceania & Polynesia	53	0.4	0.6	0.6	1.9	0.5
Asia	559	5.7	28.6	70.6	6.1	9.6
Africa	244	6.8	20.6	27.0	11.2	1.6
Centr. Am. & Carib.	43	3.1	7.3	5.7	24.1	1.8
South America	129	9.7	30.3	30.6	31.0	8.0
Developed	591	2.7	4.8	7.9	1.3	0.9
Developing	972	25.6	87.1	134.3	11.6	21.0
World	1563	28.3	91.9	142.2	7.7	21.9
Brazil	66	5.0	19.6	18.0	37.4	6.7
India	167	0.7	2.9	8.1	2.1	4.8
China	139	1.6	4.1	11.1	4.1	1.2
Thailand	19	0.1	0.6	6.3	3.0	1.0
Pakistan	21	0.1	0.5	1.1	2.5	1.0

Source: Fisher *et al.*, 2008, IIASA; FAOSTAT, 2008; calculation by the authors. Suitability classes are mutually exclusive, i.e. do not overlap.

Table 9. Suitability of unprotected grass/scrub/wood land for rain-fed sugarcane production.

	Unprotected grass/scrub/ wood land	Land potentially suitable, of which			VS+S in grass & wood land
		Very suitable	Suitable	Moderately suitable	
	million ha	million ha	million ha	million ha	percent
North America	566	1.1	2.1	3.7	0.6
Europe & Russia	666	0	0	0.0	0.0
Oceania & Polynesia	519	0.4	1.6	3.2	0.4
Asia	699	1.3	5.7	22.5	1.0
Africa	973	11.9	36.0	65.0	4.9
Centr. Am. & Carib.	98	1.1	2.4	3.5	3.6
South America	613	22.0	47.2	90.8	11.3
Developed	1741	1.2	2.5	4.5	0.2
Developing	2394	36.6	92.5	184.2	5.4
World	4135	37.8	95.0	188.7	3.2
Brazil	260	7.7	26.5	49.9	13.2
India	26	0.0	0.1	0.2	0.3
China	268	0.7	1.4	2.9	0.8
Thailand	12	0.0	0.1	1.2	0.6
Pakistan	14	0.0	0.0	0.0	0.0

Source: Fisher *et al.*, 2008; calculation by authors. Suitability classes are mutually exclusive, i.e. do not overlap.

In total, globally some 3.2 billion hectares of land are classified as unprotected forests, of which 7.3 percent were regarded as very suitable (49 million hectares) or as suitable (some 185 million hectares; see Table 10) for rain-fed sugarcane cultivation. Of the suitable extents in both of these prospective suitability classes, Africa and South America contribute about 85 percent of the total.

2.4. Sustainability of land use changes

Sugarcane is widely accepted as one of the most promising – economically and with regard to greenhouse gas saving potential – bioenergy feedstock options currently available. For instance, the fossil energy ratio (output biofuel energy per unit of fossil fuel input energy) of sugarcane ethanol was 9.3 in 2006 and is projected to reach 11.6 by 2020 with

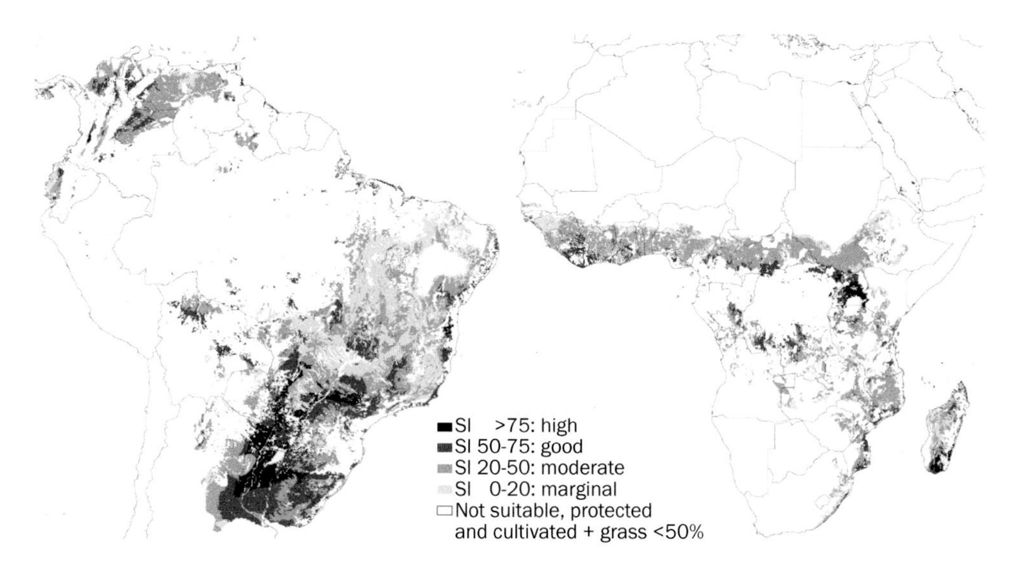

Figure 11. Suitability of current cultivated land and grass, scrub, woodland areas for rain-fed sugarcane production. Source: Fisher *et al.*, 2008.

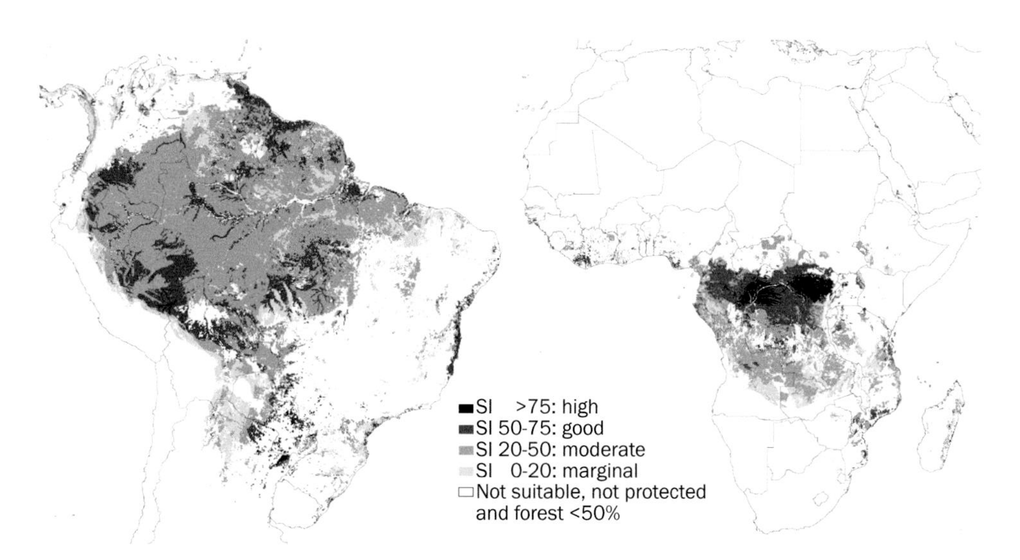

Figure 12. Hot spots of suitability of forest land for rain-fed sugarcane production. Source: Fisher *et al.*, 2008; calculation by authors.

the implementation of commercial technologies already available (Macedo *et al.*, 2008). In comparison, as reviewed by Goldemberg (2007), fossil energy ratio is 10.0 for cellulose ethanol in the United States, 2.1 for sugar beet in Europe and 1.4 for maize ethanol in the United States. The energy and greenhouse gas balance of sugarcane compares very favorably

Table 10. Suitability of unprotected forest land for rain-fed sugarcane production.

	Unprotected forest	Land potentially suitable, of which			VS+S in unprotected forest
		Very suitable	Suitable	Moderately suitable	
	million ha	million ha	million ha	million ha	percent
North America	496	3.1	8699	16.1	2.4
Europe & Russia	910	0	0	0.0	0.0
Oceania & Polynesia	121	0.8	4.6	8.2	4.5
Asia	476	1.7	10.5	41.4	2.6
Africa	444	28.0	79.5	81.4	24.2
Centr. Am. & Carib.	81	1.9	3.7	5.2	6.9
South America	694	13.1	78.2	266.9	13.2
Developed	1516	3.5	10.2	18.0	0.9
Developing	1706	45.2	175.0	401.2	12.9
World	3222	48.7	185.2	419.2	7.3
Brazil	414	4.4	45.0	174.8	11.9
India	61	0.3	0.6	2.0	1.4
China	158	0.5	1.2	2.7	1.1
Thailand	9	0.0	0.1	1.3	0.6
Pakistan	2	0.0	0.0	0.1	0.8

Source: Fisher *et al.*, 2008; calculation by authors. Suitability classes are mutually exclusive, i.e. do not overlap.

with other first generation biofuels; as reviewed in several studies, bioethanol based on sugarcane can achieve greenhouse gas reductions of more than 80% compared to fossil fuel use (e.g. Macedo (2002); Macedo *et al.* (2004); De Oliveira *et al.* (2005)).

The rapid further expansion of sugarcane areas forecasted for Brazil is expected to continue at the expense of current crop land and extensively managed pastoral land in the Cerrado region. This expansion may directly or indirectly affect parts of the Cerrado area with native vegetation and unprotected forest where biophysical, infrastructural and socio-economic conditions are favorable for sugarcane cultivation. Most threatened are those lands adjacent to current production areas. Environmental consequences of sugarcane expansion might range from quite acceptable (conversion of crop land and managed pastures) to very negative where sugarcane expands directly or indirectly in unprotected areas, which still have native vegetation with high bio-diversity or into unprotected native forest areas. Apart from the question, which land will be converted, environmental impacts will be molded by agricultural and industrial technologies applied in newly converted areas.

Current concerns regarding sustainable expansion of the sugarcane industry in Brazil (see Box 2) have been recently investigated (Goldemberg *et al.*, 2008; Martinelli and Filoso, 2008; Smeets *et al.* 2008).

Pressure on native ecosystems and threats to biodiversity can be avoided by effective environmental regulation and control and by implementation of agricultural policies supporting intensification of production. Increasing demand for food and livestock products will require replacement of the land converted to sugarcane, leading to substantial shifts of crop land and pastures to other regions, causing pressure on the ecosystems there. Such indirect land use changes would negatively affect the greenhouse gas efficiency of sugarcane production.

So far sugarcane in Brazil has mostly intruded in the cultivated and pasture areas of São Paulo State. For this state, the estimated remaining area of pastures, of which many are bordering on the sugarcane production expansion front, is 7.6 million ha (IEA, 2007). In the entire Cerrado region (205 million ha) there are currently about 54 million ha of these pastures (Ministério do Meio Ambiente, 2007).

Assuming that cultivated pastures will continue to be converted into sugarcane and that on top of this, demand for livestock products further increases, substantially higher stocking rates will be required. This implies adoption of new technologies (Corsi, 2004) for intensification of pastoral management (e.g. use of fertilizers, rotational grazing) with consequent increases of agro-chemical inputs, production costs and greenhouse gas emissions. The remaining 124 million ha of Cerrado with native vegetation (see Table 7), which are susceptible to loss

Box 2. What are key concerns and environmental issues with sugarcane expansion?

- Deforestation and habitat loss.
- Land competition with food and feed production.
- Indirect effects of land conversion because of strong expansion of sugarcane production out-competing other crop and livestock activities, which in turn encroach on natural habitats.
- Water pollution and eutrophication.
- Soil erosion and soil compaction (mainly during land preparation and early growth phases when soil is barren combined with sub-optimal tillage methods and relative high rainfall and the use of steep slopes).
- Air pollution (mainly through burning of sugarcane before harvest)
- Possible extensive use of transgenic sugarcane types

of bio-diversity and land degradation are an imminent target for sugarcane expansion and needs therefore serious attention. Expansion of protected areas, zero deforestation policies for native forest land as well as reforestation of already deforested areas are important elements of a sustainable agricultural development (Machado *et al.*, 2004; Durigan *et al.*, 2007). Currently, less than 6% of the Cerrado region is legally protected. A share of 20% of natural vegetation is required as a 'legal reserve' by the Brazilian Forest Code in this region, in comparison to 80% in the Amazon rainforest (Conservation International, 2008; Klink and Machado, 2005).

The use of genetically modified sugarcane, with associated risks of impacting biodiversity or becoming invasive in natural habitats, has been identified as an additional area of concern for future expansion of sustainable sugarcane production (Smeets *et al.*, 2008). The sequencing of sugarcane genes and development of transgenic varieties has been pursued in Brazil as a means of conferring disease resistance, stress tolerance and efficiency of nutrient use in the plant, which could contribute to sustainable expansion in the future (Cardoso Costa *et al.* 2006). The country has a well-established research in the biotechnology field with reported successes in developing disease and herbicide resistant agricultural and horticultural crops. Although potential benefits are high, there is still a lack of understanding of the potential impacts of genetically modified organisms on environmental parameters (Smeets *et al.*, 2008), which prompted the removal of permits for commercial trials with transgenic sugarcane after public concerns.

Pollution problems require strict enforcement of legislation and inspection of agricultural and industrial activities. Strict regulation and control of the disposal of nutrient-rich waste from industrial processes (e.g. vinasse) is required to avoid deterioration of water quality near production areas (Gunkel *et al.*, 2007). Recycling of byproducts of sugarcane in the fields reduces chemical fertilizers application rates; however, there is a risk of excess application in particular at close distance to the processing plants (Smeets *et al.*, 2008).

Various technologies have been identified for immediate increases in the efficiency and sustainability of current and future sugarcane mills, e.g. reducing water consumption with closure of water-processing circuits and the use of bagasse (fibrous residue left after cane milling) to generate electricity, improving the energy balance of ethanol production; as well as in production and harvesting processes. Air pollution caused by sugarcane burning can be effectively avoided by the adoption of mechanized harvesting. In São Paulo, where more than one third of the area of sugarcane is already harvested mechanically (Conab, 2008b), a schedule of phasing out burning is in place. Targets are that by 2020 all land with slopes <12% and by 2030 all the sugarcane land should harvested mechanically (Smeets *et al.*, 2008). These authors also indicate that high investment requirements and difficulties with mechanization on, for example steep land, increase the risks of the full implementation of mechanized harvest. An additional challenge are the social consequences of mechanical harvesting because of the significant losses of jobs, i.e. currently 80 workers would be replaced by one mechanical harvester (Conab, 2008a). In 2007, about three quarters of the Brazilian sugarcane area was still manually harvested and some 300,000 workers depend for their livelihood on manual cutting of cane. The pace of introduction of mechanized harvesting will therefore be affected by the cost/benefit of substituting manual labor and on suitable socio-economic conditions to reallocate the current contingent of sugarcane-cutting workers.

Adequate know-how and well developed technology is available to achieve sustainable sugarcane production and expansion (Goldemberg *et al.*, 2008). However, the adoption of new technologies requires a favorable economic and political environment that facilitates investments in clean technologies. While Brazil has accumulated considerable experience on sustainable sugarcane production through its PROALCOOL program, it will be critical to share and transfer this knowledge and ensure application of new technologies and of 'best practices' in other regions of the Americas, Asia and especially Africa, where large expansion potentials may materialize quickly due to the current urgency to develop bioenergy resources.

References

Cançado, J.E.D., P.H.N. Saldiva, L.A.A. Pereira, L.B.L.S. Lara, P. Artaxo, L.A. Martinelli, M.A. Arbex, A. Zanobetti and A.L.F. Bragal, 2006. The Impact of Sugarcane-Burning Emissions on the Respiratory System of Children and the Elderly. Environmental Health Perspectives 114: 725-729.

Cardoso Costa, M.G., A. Xavier and W. Campos Otoni, 2006. Horticultural biotechnology in Brazil. Acta Hort. 725: 63-72.

Conab, 2008a. Companhia Nacional de Abastecimento. Acompanhamento da safra brasileira. Abril 2008. Available at: http://www.conab.gov.br/conabweb/index.php?PAG=131. Retrieved 13 August, 2008.

Conab, 2008b. Companhia Nacional de Abastecimento. Perfil do Setor de Açúcar e de Alcool no Brasil. Available at: http://www.conab.gov.br/conabweb/index.php?PAG=131. Retrieved 13 August, 2008.

Conab, 2008c. Companhia Nacional de Abastecimento. Séries Históricas. Available at: http://www.conab.gov.br.

Conservation International, 2008. Biodiversity hot spots. Conservation International. Washington DC, USA. Available at: http://www.biodiversityhotspots.org/xp/hotspots/cerrado/Pages/default.aspx.

Corbi, J.J., S.T. Strixino, A. Santos and M. Grande, 2006. Diagnóstico ambiental de metais e organoclorados em córregos adjacentes a áreas de cultivo de cana-de-açúcar (Estado de São Paulo, Brasil). Quimica Nova 29: 61-65.

Corsi, M., 2004. Impact of grazing management on the productivity of tropical grasses. International Grassland Congress. Available at: http://www.internationalgrasslands.org/publications/pdfs/tema22_2.pdf

De Oliveira, M., B. Vaughan and E. Rykiel Jr., 2005. Ethanol as a Fuel: Energy, Carbon Dioxide Balances and Ecological Footprint. Bioscience 55: 593-602.

Durigan, G., M.F. d. Siqueira and G.A.D.C. Franco, 2007. Threats to the Cerrado remnants of the state of São Paulo, Brazil. Scientia Agricola 64: 355-363.

EDC, 2000. Global Land Cover Charateristics Database version.2.0 (http://edcwww.cr.usgs.gov).

FAO, 1987. 1948-1985 World Crop and Livestock Statistics. Food and Agriculture Organization of the United Nations, Rome, Italy.

FAO, 2001. Global Forest Resources Assessment 2000, FAO, Rome, Italy.

FAO/IIASA/ISRIC/ISSCAS/JRC, 2008. Harmonized World Soil Database (version 1.0). FAO, Rome, Italy and IIASA, Laxenburg, Austria.

FAOSTAT, 2008. Production. Available at: http://faostat.fao.org.

Fearnside, P.M., 2005. Deforestation in Brazilian Amazonia: History, rates, and Consequences. Conservation Biology 19: 680-688.

Fischer, G., M. Shah and H. van Velthuizen, 2002. Climate Change and Agricultural Vulnerability. Special Report as contribution to the World Summit on Sustainable Development, Johannesburg 2002. International Institute for Applied Systems Analysis, Laxenburg, Austria. pp 152.

Fischer, G., M. Shah, F.N. Tubiello and H van Velhuizen, 2005. Socio-economic and climate change impacts on agriculture: an integrated assessment, 1990–2080. Philosophical Transactions of the Royal Society B: Biological Sciences 360: 2067-2083.

Fischer, G., F.N. Tubiello, H. van Velthuizen and D. Wiberg, 2006. Climate change impacts on irrigation water requirements: Effects of mitigation, 1990–2080. Technological Forecasting & Social Change 74: 1083-1107.

Fischer, G., F. Nachtergaele, S. Prieler, E.Teixeira, H.T. van Velthuizen, L. Verelst, D. Wiberg, 2008. Global Agro-ecological Zones Assessment for Agriculture (GAEZ 2008). IIASA, Laxenburg, Austria and FAO, Rome, Italy.

Florestar, 2005. Florestar Estatistico - Fundo de Desenvolvimento Florestal. Cobertura Florestal em São Paulo. Available at: http://www.floresta.org.br/index.php?interna=estatisticas/florestarestatistico&grupo=3. Retrieved 13 August, 2008.

Goldemberg, J., 2006. The ethanol program in Brazil. Environmental Research Letters 1: 5.

Goldemberg, J., 2007. Ethanol for a Sustainable Energy Future. Science 315: 808-810.

Goldemberg, J., S.T. Coelho and P. Guardabassi, 2008. The sustainability of ethanol production from sugarcane. Energy Policy 36: 2086-2097.

Grübler, A., B. O'Neill, K. Riahi, V. Chirkov, A. Goujon, P. Kolp, I. Prommer and E. Slentoe, 2006. Regional, national and spatially explicit scenarios of demographic and economic change based on SRES. Technological Forecasting & Social Change 74: 980-1029.

Gunkel, G., J. Kosmol, M. Sobral, H. Rohn, S. Montenegro and J. Aureliano, 2007. Sugarcane Industry as a Source of Water Pollution - Case Study on the Situation in Ipojuca River, Pernambuco, Brazil. Water Air Soil Pollution 180: 261-269.

IEA, 2007. Instituto de Economia Agrícola. Área e Produção dos Principais Produtos da Agropecuária-São Paulo. Available at: http://www.iea.sp.gov.br/out/banco/menu.php. Retrieved 12 August, 2008.

IFPRI, 2002. Global agricultural extent: Reinterpretation of global land cover characteristics database (GLCCD v. 2.0), EROS data center (EDC), 2000. International Food Policy Research Institute, Washington DC, USA.

Klink, C.A. and R.B. Machado, 2005. Conservation of the Brazilian Cerrado. Conservation Biology 19: 707-713.

Licht, F.O., 2007. World Ethanol and Biofuels Report, Oct 23, 2007.

Licht, F.O., 2008. World Ethanol and Biofuels Report, May 5, 2008.

Macedo, I.C., J.E.A. Seabra and J.E.A.R. Silva, 2008. Green house gases emissions in the production and use of ethanol from sugarcane in Brazil: The 2005/2006 averages and a prediction for 2020. Biomass and Bioenergy 32: 582-595.

Macedo, I., 2002. Energia da Cana no Brasil. Unicamp – São Paulo, 2002. Available at: http://www.cgu.rei.unicamp.br/energia2020/papers/paper_Macedo.pdf

Macedo, I., M.R.L.V. Leal and J.E.A.R.d. Silva, 2004. Greenhouse gas emissions in the production and use of ethanol in Brazil: present situation (2002). Secretaria de Meio-Ambiente do Estado de São Paulo, Brazil.

Machado, R.B., M.B. Ramos Neto, P. Pereira, E. Caldas, D. Goncalves, N. Santos, K. Tabor and M. Steiniger, 2004. Estimativa de perda do Cerrado brasileiro. Conservation International do Brazil, Brasilia, DF, Brazil.

Martinelli, L.A. and S. Filoso, 2008. Expansion of sugarcane ethanol production in Brazil: environmental and social challenges. Ecological Applications 18: 885-898.

Ministério do Meio Ambiente, 2007. Mapeamento de Cobertura Vegetal do Bioma Cerrado. Relatório Final. Brasília/DF. Junho 2007. Available at: http://www.mma.gov.br/index.php?ido=conteudo.monta&idEstrutura=72&idMenu=3813&idConteudo=5978.

Mitchell, T.D. and P.D. Jones, 2005. An improved method of constructing a database of monthly climate observations and associated high resolution grids. Climatic Research Unit, School of Environmental Sciences, University of East Anglia, Norwich, United Kingdom.

Myers, N., R.A. Mittermeier, C.G. Mittermeier, G.A.B. da Fonseca and J. Kent, 2000. Biodiversity hotspots for conservation priorities. Nature 403: 853-858.

Naseri, A.A., S. Jafari and M. Alimohammadi, 2007. Soil Compaction Due to Sugarcane (Saccharum officinarum) Mechanical Harvesting and the Effects of Subsoiling on the Improvement of Soil Physical Properties. Journal of Applied Sciences 7: 3639-3648.

New, M., D. Lister, M. Hulme and I. Makin, 2002. A high-resolution data set of surface climate over global land areas. Climate Research 21: 1-25.

Oliveira, J.C.M., C.P.M. Vaz and K. Reichardt, 1995. Efeito do cultivo continuo da cana de acucar em propriedades fisicas de um latossolo vermelho escuro. Scientia Agricola 52: 50-55.

Politano, W. and T.C.T. Pissarra, 2005. Avaliação por fotointerpretação das áreas de abrangência dos diferentes estados da erosão acelerada do solo em canaviais e pomares de citros. Engenharia Agrícola 25: 242-252.

Ribeiro, H., 2008. Sugarcane burning in Brazil: respiratory health effects. Rev. Saúde Pública 42: 1-6.

Saint, W.S., 1982. Farming for energy: social options under Brazil's National Alcohol Programme. World Development 10: 223-238.

Sano, E.E., R. Rosa, J.L.S. Brito and L.G. Ferreira, 2008. Mapeamento semidetalhado do uso da terra do Bioma Cerrado. Pesquisa Agropecuária Brasileira 43: 153-156.

Scharlemann, J.P.W. and W.F. Laurance, 2008. Environmental Science: How Green Are Biofuels? Science 319: 43-44.

Silva, A.M.d., M.A. Nalon, F.J.d.N. Kronka, C.A. Alvares, P.B.d. Camargo and L.A. Martinelli, 2007. Historical land-cover/use in different slope and riparian buffer zones in watersheds of the state of São Paulo, Brazil. Scientia Agricola 64: 325-335.

Silva, D.M.L.d., P.B.d. Camargo, L.A. Martinelli, F.M. Lanças, J.S.S. Pinto and W.E.P. Avelar, 2008. Organochlorine pesticides in Piracicaba river basin (São Paulo/Brazil): a survey of sediment, bivalve and fish. Quimica Nova 31: 214-219.

Smeets, E., M. Junginger, A. Faaij, A. Walter, P. Dolzan and W. Turkenburg, 2008. The sustainability of Brazilian ethanol-An assessment of the possibilities of certified production. Biomass and Bioenergy 32: 781-813.

Sparovek, G. and E. Schnug, 2001. Temporal Erosion-Induced Soil Degradation and Yield Loss. Soil Sci Soc Am J, 65: 1479-1486.

Tercil, E.T., A.M.d.P. Peres, M.T.M. Peres, S.N.R. Guedes, P.F.A. Shikida and A.M.C.J. Corrêa, 2007. Os Mercados de Terra e trabalho na (re)estruturação da categoria social dos fornecedores de cana do Estado de São Paulo: análise de dados de campo. REDES, Santa Cruz do Sul 12: 142-167.

Tominaga, T.T., F.A.M. Cássaro, O.O.S. Bacchi, K. Reichardt, J.C.M. Oliveira and L.C. Timm, 2002. Variability of soil water content and bulk density in a sugarcane field. Australian Journal of Soil Research 40: 605-614.

Tubiello, F.N. and G. Fischer, 2006. Reducing climate change impacts on agriculture: Global and regional effects of mitigation, 2000–2080. Technological Forecasting & Social Change 74: 1030-1056.

Chapter 3
Prospects of the sugarcane expansion in Brazil: impacts on direct and indirect land use changes

André Meloni Nassar, Bernardo F.T. Rudorff, Laura Barcellos Antoniazzi, Daniel Alves de Aguiar, Miriam Rumenos Piedade Bacchi and Marcos Adami

1. Introduction

Sugarcane has been an important crop since the initial colonization period of Brazil and is nowadays expanding considerably its cultivated area, particularly due to strong ethanol demand. Ethanol demand has been increasing in the internal market since 2003 - due to the expansion of the flex-fuel car fleet - and is also facing good perspectives in the international market. From 2000 to 2007 the cultivated sugarcane area increased by about 3 million ha, reaching about 7.9 million ha based on information from IBGE (2008a). The South-Central region was responsible for 95.7% of this total growth.

Sustainability of agricultural based biofuels has turned into a central question once the use of biofuels with the aim to reduce greenhouses gases' (GHG) emissions increases. The full life cycle analysis of the production process of every feedstock, based on carbon equivalent emissions, is the essential measure for assessing the sustainability of biofuels.

The agricultural component of the biofuel production is, therefore, a key variable for determining the avoided carbon emissions. Agricultural products are, by its nature, large land users. Crops - annual and permanent - and cattle - for dairy and beef - occupy about 77 and 172 million ha, respectively, in Brazil (IBGE, 2008b). Land use changes due to the competition between crops and cattle may raise concerns in terms of GHG emissions and it becomes even more important when land with natural vegetation (mainly forests and Cerrado) is converted into cattle raising or agricultural production. There is no recognized and unquestionable methodology to measure the amount of deforestation caused by agricultural expansion. However, the amount of land allocated to pastures and crops in the frontier are indicators that both processes are correlated.

Given that sustainability of Brazilian ethanol is intrinsically associated to the sugarcane expansion's effects on land use changes, this paper aims to analyze past and expected sugarcane expansion in Brazil and to understand the land use change process. Competition between food and biofuel increases the importance of this issue and has been adding also social and economic concerns about land use change caused by biofuels' expansion. Different opinions from many international organizations, national governments, NGOs and researchers are putting this debate in the centre of media and public opinion worldwide.

Considering that this debate has not yet been explored in depth in Brazil, this paper aims to support these discussions with technical and scientific arguments.

Land use change, as a consequence of the expansion of agricultural production as well as due to the competition for land among agricultural activities, is an issue under development in Brazil in terms of economic analysis and modeling. With exception of the analysis focused on land use changes related to deforestation in the Brazilian Amazon, which is well monitored by Brazilian government agencies and environmentalists non-profit organizations, there is not regular monitoring of the conversion of natural landscapes into agricultural uses. Furthermore, there is a lack of economic models that are able to explain and predict land allocation and land use change as a consequence of the dynamics of crops and pasture land. This paper is a result of one of the initiatives under development in Brazil in order to clarify this issue.

However, the complexity associated to measure land use change in the context of assessing biofuel's carbon life cycle is largely related to the extension of the concept. Two approaches are under scrutiny: direct land use change (LUC); and indirect land use change (ILUC). The objectives of the present study are to measure and evaluate direct changes of land use caused by the sugarcane expansion over the last years as well as the consequences of future expected expansion. Land use changes are measured in terms of crops and pasture directly displaced by the sugarcane expansion. The study also aims to discuss indirect land use change related to Brazilian sugarcane expansion. Information and data are presented in order to evaluate effect-cause relationships between sugarcane and other agricultural expansion areas.

The measurement of land use changes as a consequence of agricultural production expansion, looking to the past and forecasting the future, is a very dynamic and complex process. This paper searches for support on different methodologies to understand this process: (1) when measuring the past land use change, primary data based on remote sensing images and environmental licensing reports as well as secondary data based on planted and harvested area are used; (2) with respect of projections of land allocated to sugarcane, a partial equilibrium model based on profitability and demand/supply responses to price variations is developed.

The paper is organized as follows. Section two introduces the discussion of the dynamics of the sugarcane expansion in Brazil and shows that the expansion is highly concentrated in the South-Central region. Section three presents the different methodologies used in this study to measure land use change due to past and future expansion of sugarcane. Section four is also divided in two perspectives, past and future expansion, and presents the results of the assessment of the sugarcane expansion and the consequent crop and pasture displacement. Conclusions and recommendations are presented in section five.

2. The dynamics of sugarcane expansion in Brazil

Before discussing land use changes caused by the sugarcane expansion, it is important to know how the sugarcane crop is spatially distributed in Brazil. Almost all of the sugarcane in Brazil is produced in the South-Central and Northeast regions. These two regions are considered separately due to its different harvest seasons: the first is from April to October and the second from November to March.

Based on data from the Canasat Project for the states of São Paulo, Minas Gerais, Paraná, Mato Grosso, Mato Grosso do Sul and Goiás and from the IBGE (Brazilian Institute of Geography and Statistics) for all other states, it is estimated that in 2008 the cultivated area in the South-Central region was 7.4 million ha (85.0%) and in the Northeast region was 1.3 million ha (14.7%).

Figure 1 presents the evolution of sugarcane area for three regions: South-Central (comprising its six most important states)[2], Northeast and all the other states. According to the statistics of IBGE the Northeast region has a relative steady sugarcane area, presenting a mean annual growth rate of only 2%; while in the South-Central region the annual growth rate was 16% over the last four years being responsible for 95.4% of the total sugarcane area expansion in Brazil from 2005 to 2008. During this period the sugarcane area in Brazil expanded by an annual rate of 13% (2.6 million ha) going from 6.1 to 8.7 million ha (Figure 1).

São Paulo is the most important state for sugarcane, representing 55.7% of the total sugarcane area in Brazil in 2008. The four states with the largest sugarcane area are São Paulo, Paraná, Minas Gerais and Goiás which are responsible for 75.2% of total sugarcane area in Brazil. Coincidently, these states plus Mato Grosso and Mato Grosso do Sul have experienced the greatest sugarcane expansion area over the last years. A new and promising region for sugarcane is located in the states of Maranhão, Piauí and Tocantins, in the Cerrado biome, commonly known in Brazil as the MAPITO region; however, in 2008 these states were responsible for only 0.25% of the cultivated sugarcane area in Brazil.

South-Central, including MAPITO region, is here called Expanded South-Central and is considered to be a relevant region for sugarcane expansion analysis. The sugarcane area in the Expanded South-Central in 2008 was 7.5 million hectares (84% of total area) and represents 97% of the total sugarcane expansion. In all other Brazilian states not included in the Expanded South-Central and Northeast regions, which accounts for 3% of the total sugarcane area, a reduction of 13.7 thousand hectares was observed from 2005 to 2008.

[2] Although South-Central region is commonly defined as the states in South, Southeast and Centre-West political regions, this paper refers to the South-Central as a region comprising São Paulo, Minas Gerais, Paraná, Goiás, Mato Grosso do Sul and Mato Grosso. This definition is in line with the satellite images monitored by the Canasat Project, due to the fact they are the most important states in terms of sugarcane expansion.

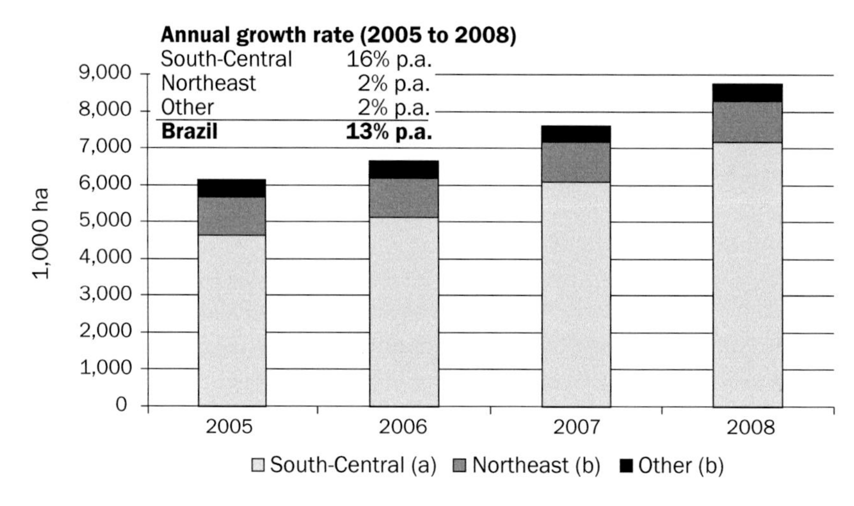

Figure 1. Sugarcane area cultivated in Brazil according to production regions (2005 to 2008). Sources: (a) Canasat/INPE, comprising São Paulo, Minas Gerais, Paraná, Goiás, Mato Grosso and Mato Grosso do Sul; (b) PAM/IBGE (2005 and 2006) and LSPA/IBGE (2007 and 2008).

Therefore, for this work the evaluation of the conversion of land use and occupation for sugarcane is restricted to a reduced South-Central region that comprises the states of São Paulo, Minas Gerais, Goiás, Paraná, Mato Grosso do Sul and Mato Grosso (or South-Central region minus the states of Rio de Janeiro, Espírito Santo, Santa Catarina and Rio Grande do Sul)[3] in Sections 3.1 and 4.1 and these states plus Maranhão, Piauí, Tocantins and Bahia in Sections 3.2 and 4.2. As a means of illustration, a detailed visualization of the sugarcane distribution in the South-Central region is presented in Figure 2.

3. Methodology

This papers divides the analysis of land use changes (LUC) caused by sugarcane expansion basically in observed LUC (past trend) and projected LUC (future trend). Three different methods were used to estimate past land use dynamics, and another one to project future trend. For observed LUC and sugarcane expansion, we used the information extracted from remote sensing images, secondary data by IBGE, and field research through environmental licensing studies. The satellite image analysis was carried out for São Paulo, Minas Gerais, Paraná, Goiás, Mato Grosso and Mato Grosso do Sul. Using IBGE data, all these states plus Tocantins, Maranhão, Piauí and Bahia were analyzed due to their potentiality for future expansion of agricultural area in Brazil. For the field research, the analyzed states were São

[3] The states of Rio de Janeiro, Espírito Santo, Santa Catarina and Rio Grande do Sul are part of the South-Central region but represent only 3.2% of the cultivated sugarcane area in 2008 and have not shown relevant sugarcane expansion over the last years (Table 3).

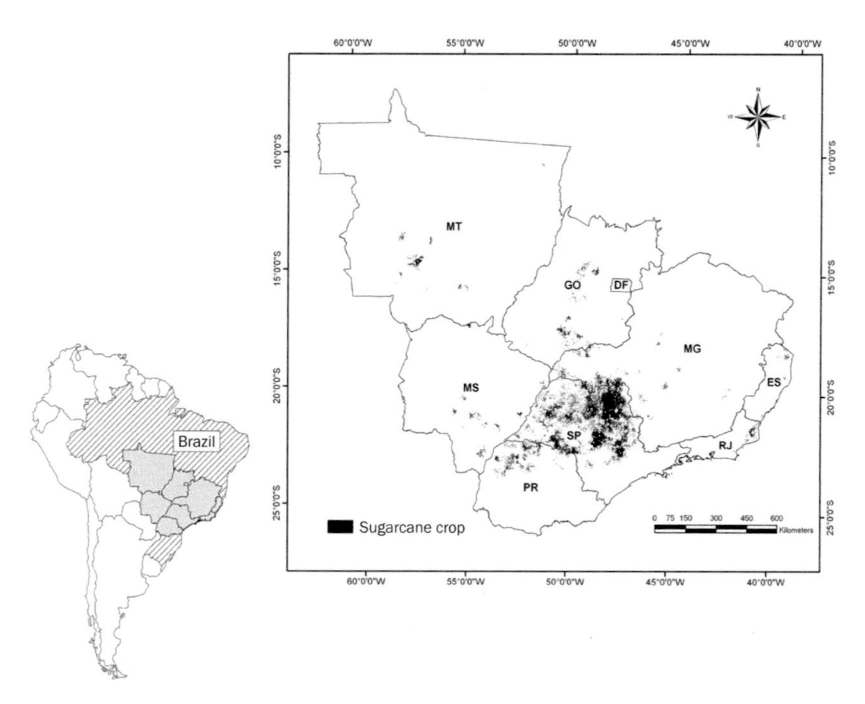

Figure 2. Spatial distribution of sugarcane crop in 2007 in the South-Central region of Brazil. Source: INPE (www.dsr.inpe.br/canasat/).

Paulo, Minas Gerais, Mato Grosso, Mato Grosso do Sul, Goiás and Tocantins, capturing both past and near future trends. A deeper description of each method follows below.

It is important to clarify the available sources of data regarding sugarcane area in Brazil, since different ones are used for different purposes here. The Brazilian Institute for Geography and Statistics (IBGE) is the Brazilian official organization which provides data on crops area, production, yield, and several others variables. For estimation of planted and harvested sugarcane area, two IBGE databases can be used: one from the systematical survey on agricultural production (LSPA – Levantamento Sistemático da Produção Agrícola) and the other from the agricultural production by municipality (PAM – Produção Agrícola Municipal). While the former includes all area occupied with sugarcane – which consists of areas to be harvested and new areas, to be harvested only next year – the latter just includes harvested area in a certain year[4]. PAM data is available for all geographic scales, from 1990 to 2006 while LSPA data are forecasts from previous and current years subjected to change.

[4] Sugarcane crop is harvested yearly for 5 or 6 consecutive years. In general, after that period the sugarcane field is renewed and generally rotated for soil improvement during one season with a crop from the *Leguminosae* family. Planted area (or total land occupied by sugarcane), therefore, is the area available for harvest of sugarcane plus the area of sugarcane that is being renewed.

Another source of data on sugarcane planted area is the National Food Supply Company (CONAB). CONAB data, which are presented in a crop assessment reports, are not used in this study.

Sugarcane planted area is also provided on the Canasat Project, coordinated by the National Institute for Space Research (INPE). Canasat Project monitors the most important producing states and estimates the planted area from remote sensing satellite images. A summary of data available in Brazil is presented in the Table 1.

Table 1. Sugarcane area in Brazil (sources of data available).

Source	Data gathering	Period available	Data presented	Aggregation level
IBGE	LSPA survey	1990-2006	planted area and harvested area	state (Brazil)
	PAM survey	1990-2006	harvested area	municipality (Brazil)
CONAB	crop assessment	2005, 2007-2008	planted area	state (Brazil)
INPE/Canasat	satellite images	2003/2005 - 2008	planted area	municipality (South-Centre)

3.1. Measuring the LUC using remote sensing images

Remote sensing satellite images from the Earth surface are an important source of information to evaluate the fast land use changes observed by the dynamic agricultural activity. Brazil, through its National Institute for Space Research (INPE), acquires remote sensing images from both Landsat and CBERS satellites since 1973 and 1999, respectively. In 2003, INPE started the Canasat Project together with UNICA (Sugarcane Industry Association), CEPEA (Center for Advanced Studies on Applied Economics) and CTC (Center of Sugarcane Technology) to map the cultivated sugarcane area in the South-Central region of Brazil using remote sensing images (www.dsr.inpe.br/canasat/).

The mapping began in São Paulo State in 2003 and in 2005 it was extended to the states of Minas Gerais, Goiás, Mato Grosso do Sul and Mato Grosso where sugarcane production has been most intensified over the last years. In order to obtain an accurate thematic map, multitemporal images were acquired at specific periods to correctly identify the sugarcane fields and to clearly distinguishing them from other targets (Rudorff and Sugawara, 2007). The mapping procedure was mainly performed through visual interpretation on the computer screen using the SPRING software (www.dpi.inpe.br/spring/), which is a Geographic Information System (GIS) with digital image processing capabilities.

Results from the thematic maps of the Canasat Project presented in Section 4.1 refer to the total cultivated sugarcane area which includes the fields being renewed with the 18 months sugarcane plant, the fields planted in new areas (expansion) and the fields of sugarcane ratoons (Sugawara *et al.*, 2008). Prior land use identification for the expanded sugarcane plantations in each year was carried out using remote sensing images acquired before the land use change to sugarcane. This evaluation was accomplished in São Paulo for the years of 2005 to 2008, and for the years of 2007 and 2008 for the States of Minas Gerais, Goiás, Paraná, Mato Grosso do Sul and Mato Grosso.

Four classes of land use and occupation were defined, as follows: *Agriculture,* for cultivated and bare soil fields; *Pasture*, for natural and anthropogenic pasture land; *Reforestation*, for reforested areas with *Pinus* and *Eucalyptus*; and *Forest*, for riparian forests and other forests no matter the stage of succession. In São Paulo State the *Citrus* class was also considered due to its relevance in terms of land occupation and change to sugarcane. These five classes were responsible for almost all of the changes to sugarcane. Figure 3 illustrates each of these classes over some Landsat images acquired at two different dates prior to the change and one date after the change to sugarcane. Figure 3a highlights a field classified as *Agriculture.* On Date 1 (March of 2003), the field has the appearance of bare soil (medium-gray), and on Date 2 (May of 2003), it is covered with a winter crop - probably maize. On Date 3 (April of 2008), a well grown sugarcane field can be clearly identified (light-gray with well defined pathways). An example for the *Pasture* class is illustrated in Figure 3b where it appears as a mixture of different amounts of vegetation and soil (medium/light-gray). On Date 1, the vegetation amount is dominant (end of rain season) whereas on Date 2 the soil becomes dominant due to a reduction in the green vegetation amount in response to less available water to the plant (mid dry season). On Date 3, a sugarcane field can be observed in substitution to the pasture field. Figure 3c illustrates the *Citrus* class with its typical pattern on Date 1 and 2, and a sugarcane field on Date 3. Figure 3d presents a typical field for the *Forest* class (Dates 1 and 2) that was changed to sugarcane (Date 3). A field changed from the *Reforestation* class to sugarcane is illustrated in Figure 3e.

It is worth to mention that Figure 3 only illustrates, in a simplified way, part of the whole procedure used to identify the different land use classes that were displaced to sugarcane in each year over the analyzed period. In several occasions a greater number of images were necessary to clearly identify the classes that were changed to sugarcane. The SPRING software allows coupling images acquired at different dates to alternate views of the same area facilitating the visual interpretation resulting in a better extraction of the correct information registered in the coloured multispectral satellite images.

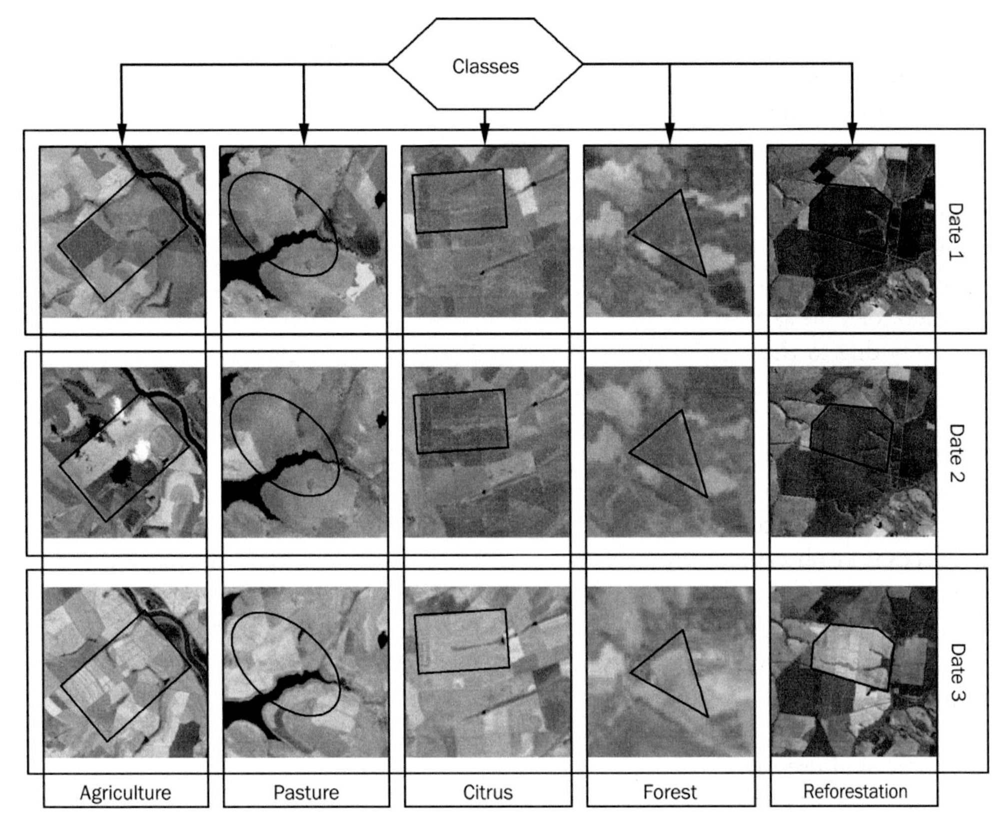

Figure 3. Different land use classes over multispectral (bands 3, 4 and 5) black and white Landsat images acquired in March of 2003 (date 1), May of 2003 (date 2), and April of 2008 (date 3).

3.2. Measuring the LUC using micro-regional secondary data

The objective of this method is to analyze the official secondary data about sugarcane and other agricultural land uses in order to verify how substitution among these uses have taken place over the last years, in different areas all over Brazil. Knowing how sugarcane expansion has occurred and how other land uses have behaved is a first step to make considerations about LUC and ILUC caused by ethanol production.

The analysis developed here was based on the Shift-share model adapted for the purposes of this study. The Shift-share model looks at the mix of activities and whether they are shifting towards or away from the area being studied (Oliveira *et al.*, 2008).

The Shift-share model decomposes the growth area of an agriculture activity in a region over a given period of time into two components: (1) growth effect, which is the part of the change attributed to the growth rate of the agriculture as a whole; and (2) an agriculture mix effect (substitution effect), which is the change in each crop share of the total cultivated area. The sum of the two effects equals to the actual change in total sugarcane area within a region over a time period.

We have used sugarcane and other crops harvested area data from IBGE, by 'micro-region', which is an aggregated number of some closely located municipalities with geographic similarities. The main advantage of using data by 'micro-region' is that it is a sufficient small unit so that land substitution is captured in a more direct way, avoiding leakages. Furthermore, the same method using data by municipality has presented some problems, probably due to the fact that municipalities have correlated productive relations and sometimes municipality data are not very accurate.

We have considered 2002 as the baseline year when the last significant sugarcane expansion began and 2006 as the last year of available data from PAM-IBGE. Since expanded sugarcane area, harvested in a certain year, was planted about 12 or 18 months before, the prior land use needs to be observed still another year before. For example, 2006 sugarcane data regards sugarcane harvested in 2006 and most likely planted in the beginning of 2005; therefore, land use prior to the 2005 sugarcane plantation should be observed in the data from 2004. Thus we have compared sugarcane expansion from 2002 to 2006 to other land occupation from 2001 to 2005.

The three land use categories are: (a) sugarcane; (b) other crops (annual and permanent crops, excluding sugarcane and second crops); and (c) pasture. Pasture land was estimated by using cattle stocking rate because data on pasture area are available only on the IBGE Agricultural Census of 1996 and 2006, while cattle herd data is available annually. Thus, stocking rate for 1996 and 2006 were calculated and an annual average growth for this period was considered. Pasture area for the analyzed years – 2001 to 2005 – were obtained dividing herd by the stocking rate. Total agricultural area is the sum of these three categories, and should represent agricultural dynamics in general. The data used for analysis was the difference between the final period and the baseline, thus positive numbers mean that there was an increase in the period while negative numbers mean that the area has decreased.

Following a logic tree of land use dynamics for the period, the 'micro-regions' were divided in six categories (Figure 4). We have only considered for analysis those 'micro-regions' where sugarcane area has increased, which means groups 2, 3 and 6. Thus expansion of sugarcane was distributed proportionally through decreased areas of pasture land and crops. When it was not possible to allocate sugarcane expansion over these land uses, it was considered not allocated over previous productive areas, meaning whether already anthropized areas not used, such as idle areas, or natural landscapes. Using shift-share terminology, sugarcane expansion over pasture and crops is considered substitution effect, while those over not productive areas are considered growth effect.

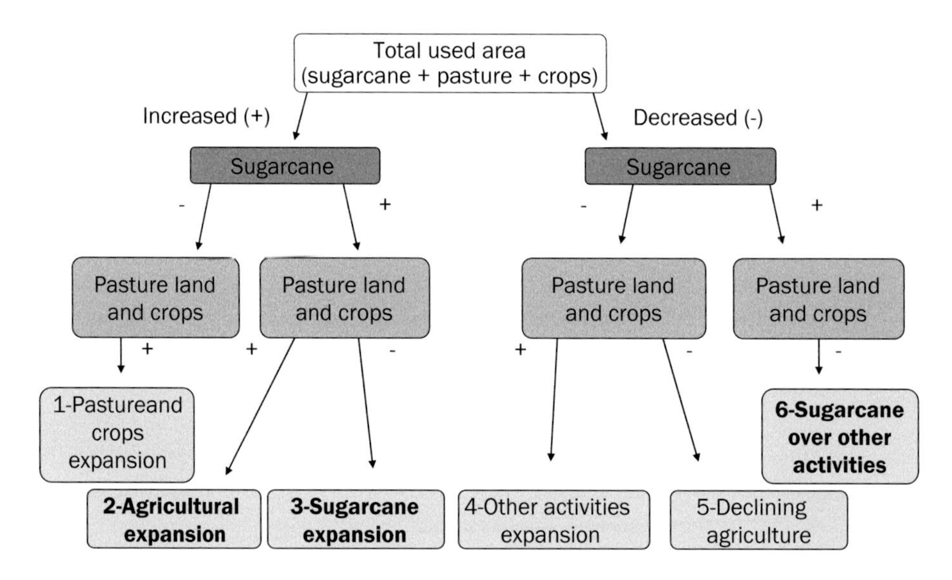

Figure 4. Land use dynamics' categories used for micro-regional secondary data analysis.

3.3. Case studies through environmental licensing reports (past and near future trend)

This third method is an empirical study which aims to collect field data from sugarcane mills in six states where the crop is under strong expansion or is expected to be in near future: São Paulo, Minas Gerais, Mato Grosso, Mato Grosso do Sul, Goiás and Tocantins[5].

Environmental Impact Assessments presented in files of environmental licenses in governmental bodies can be a useful source of information about many environmental, social and economic impacts caused by any business venture. Environmental licensing is an important instrument present in the Brazilian Environmental Policy and all sugarcane mills are required to have this license to operate. Governmental agencies responsible for issuing environmental licenses - state bodies in this case – define which type of study will be necessary for the entrepreneur to present. The most complex type of study would be the Environmental Impact Study (EIA, in Portuguese acronym) and the respective Environmental Impact Report (RIMA, which is a synthesis of EIA in a non-technical language). This study, made by the entrepreneur, contains a full characterization of the business venture, a diagnosis of the surrounding area (here including information on land use), and the impacts it will cause (natural vegetation suppression included). Even in the less complex studies, this kind of information is generally available.

[5] State of Paraná was excluded due to research restrictions imposed by the environmental authority of the state government.

Through the environmental studies used to get this license, and through the governmental agencies' database, one can obtain the exact location of mills and the economically feasible surrounding area where sugarcane is or will be cultivated. Knowing current land use in these areas by the time of the study, one can know land use prior to sugarcane plantations, as well as the type of original natural landscape when this is present.

Brazilian environmental licensing system is composed of three different phases, each of them resulting in a different license. The environmental study required by the appropriate governmental agency has to be submitted and approved to obtain the Previous License (LP, in Portuguese). The LP certifies the environmental viability of the project at that specific location, and it means the entrepreneur can ask for the Installation License (LI). Just with the LI it is legal to start the construction of the mill plant, but to initiate operation and production, the Operation License (LO) is required. The LP analysis generally takes long, because the agency must analyze deep studies, such as EIA-RIMAs, while to issue the other two licenses the approval periods are usually smaller.

By knowing which phase of the environmental licensing a mill project is ongoing or when the LO was issued, one can know when the mill will start or has started production. Furthermore, the environmental study specifies the business plan, which includes expected date to start-up. Field research was conduced between May and June, 2008.

3.4. Projecting sugarcane expansion and changes in land use

The approach used for projecting land allocation for sugarcane is based on a partial equilibrium model that is under development by the Institute for International Trade Negotiations (ICONE). The model is based on demand response to price changes and supply response to market returns (profitability) changes. National and regional prices are calculated according to a basic assumption of microeconomics: they are achieved when supply and demand prices for each coincide, generating a market equilibrium.

The model comprises 11 categories of products (sugarcane, soybean, maize, cotton, rice, dry beans, milk, beef, chicken, eggs and pork) and 6 regions: South (states of Rio Grande do Sul, Santa Catarina and Paraná), Southeast (states of São Paulo, Minas Gerais, Rio de Janeiro and Espírito Santo), Central-West Cerrados (states of Goiás, Mato Grosso do Sul and Mato Grosso under the Cerrados ecosystem), North-northeast Cerrados (states of Tocantins, Bahia, Piauí and Maranhão under the Cerrado ecosystem), Amazon Biome (states of Acre, Amazonas, Rondônia, Roraima, Amapá, Pará and Mato Grosso under the Amazon ecosystem) and Northeast Coast (states of Ceará, Rio Grande do Norte, Paraíba, Pernambuco, Alagoas and Sergipe). Estimated projections are performed on a yearly basis over a period of 10 years.

Given that the model is under development, a reduced version, comprising only the land allocation components, is used for projections presented here. Domestic demand and expected net trade are calculated exogenously, using world prices contained in the Fapri's 2008 Outlook (Fapri, 2008). Production is calculated in order to meet demand and net trade needs.

Total national area allocated for grains and sugarcane is calculated according to yields trends. In the case of pasture, total national area is the sum of regional areas. For each region, land allocation equations were specified according to the competition among crops and cattle. The allocation of land on the six regions and between product's categories is calculated on the model's land use component. The amount of land to be allocated for each activity is defined according to the required production to meet the domestic demand and net trade projections and yields trends. Allocated land is calculated by region, and crops and pasture compete according to the expected market returns per hectare. For each region and activity, specific competition equations were defined and elasticities were calculated. Land use change is measured comparing, in each region, the amount of land allocated for each activity. Absolute annual variation indicates activities that are incorporating land and the ones that are being displaced.

The competition matrix used for the specification of the land allocation regional models is described in Table 2. The competition matrix was defined according to trends in harvested area observed from 1997 to 2008 and comparing market returns per hectare. Activities with higher market returns tend to have a lower number of competitors. Historical market returns also indicate activities that are land taker and land releaser. Cattle, for example, is a typical land releaser activity and, therefore, all crops compete with pasture.

It is important to mention that the regional area allocated to pastures is calculated independently of the herd projections. Projections on regional herd and pasture must be compared to evaluate whether they are compatible with stocking rate (number of animals per hectare) trends. Although this is a limitation, because the stocking rate is not endogenous on the model, the results presented in this study are in line with the past trends in stocking rate.

Regional data, however, is too aggregate for the objective of evaluating direct land use effect of sugarcane expansion. Regional projections are breakdown by micro-regional level, in order to obtain the same level of disaggregation used in the section 3.3. Products categories are aggregated in sugarcane, pastures and grains[6]. Projected land use for each category is disaggregated according to the evolution of market share of each micro-region in the region.

[6] Grains, for the objective of the projections, comprise the following activities: soybean, maize, cotton, rice and dry beans.

Table 2. Regional land competition matrix.

Region and competing product	Cotton	Sugarcane	Soybean	Maize	Rice	Dry beans	Cattle/pasture
	Product (dependant variable)						
South		soybean maize	sugarcane maize	sugarcane soybean		soybean maize rice	all crops
Southeast		soybean maize	sugarcane maize	sugarcane soybean			all crops
Center-west Cerrados	soybean maize	soybean maize	cotton sugarcane maize	cotton sugarcane soybean	soybean		all crops
Amazon biome	soybean maize		cotton maize	cotton soybean	soybean		all crops
Northeast coastal		maize				maize	all crops
North-northeast Cerrados	soybean maize rice dry beans		cotton maize	cotton soybean	soybean	soybean maize rice	all crops

The calculation of the substitution of crops and pastures by sugarcane is executed following a similar methodology of the Shift share model presented in the section 3.2. Given that the projected period used was 2008 to 2018, sugarcane expansion is calculated as the absolute variation from 2018 to 2008 and crops and pastures expansion is calculated using a one year lag.

As well as the secondary data analysis for past expansion, the model is using harvested area rather than planted area. Therefore, the total area occupied with sugarcane is necessarily higher than the amount presented in the projections.

4. Results and discussions

The results are discussed following the same structure of the methodology section: (a) Sub-section 4.1 presents the past expansion measured through remote sensing techniques; (b) Sub-section 4.2 brings the results based on secondary data; (c) Sub-section 4.3 is devoted to the case studies; (d) Sub-section 4.4 presents the projections on land use for sugarcane; and (e) Sub-section 4.5 discusses options for analyzing ILUC.

A summary of the results obtained from the three different methodologies used in this study is presented in Table 3. Detailed results are discussed in the sub-sections.

Table 3. Land use classes converted to sugarcane: comparative results in the South-Central region (1000 ha).

	Period/measurement method		
	2002-2006 (harvested area)[1]	**2007-2008 (planted area)[2]**	**2008-2018 (harvested area)[3]**
Sugarcane expansion	1,030	2,184	3,848
Agriculture	122 (12%)	1,152 (53%)	1,455 (38%)
Pasture	793 (77%)	991 (45%)	2,369 (62%)
Other	114 (11%)	42 (2%)	24 (1%)

[1] Source: secondary data from IBGE.
[2] Source: satellite images.
[3] Source: projection model.

4.1. LUC evaluation through remote sensing images

This sub-section analysis the expanded sugarcane area harvested for the first time and is divided in three parts: (1) analysis comprising the years of 2007 and 2008 for the states of Minas Gerais, Goiás, Paraná Mato Grosso do Sul and Mato Grosso; (2) analysis comprising the period from 2005 to 2008 for the State of São Paulo; and (3) analysis comparing the years of 2007 and 2008 for the reduced South-Central region[3] (or South-Central region minus the states of Rio de Janeiro, Espírito Santo, Santa Catarina and Rio Grande do Sul).

4.1.1. Analysis for the years of 2007 and 2008 in the states of Minas Gerais, Goiás, Paraná, Mato Grosso do Sul and Mato Grosso

Figure 5 shows the results for the *Pasture, Agriculture, Reforestation* and *Forest* classes which were displaced for the expansion of sugarcane crop harvested for the first time in 2007 and 2008. Most of the expansion of sugarcane area took place over the *Agriculture* and *Pasture* classes. The *Agriculture* class registered the largest displaced area to sugarcane except for

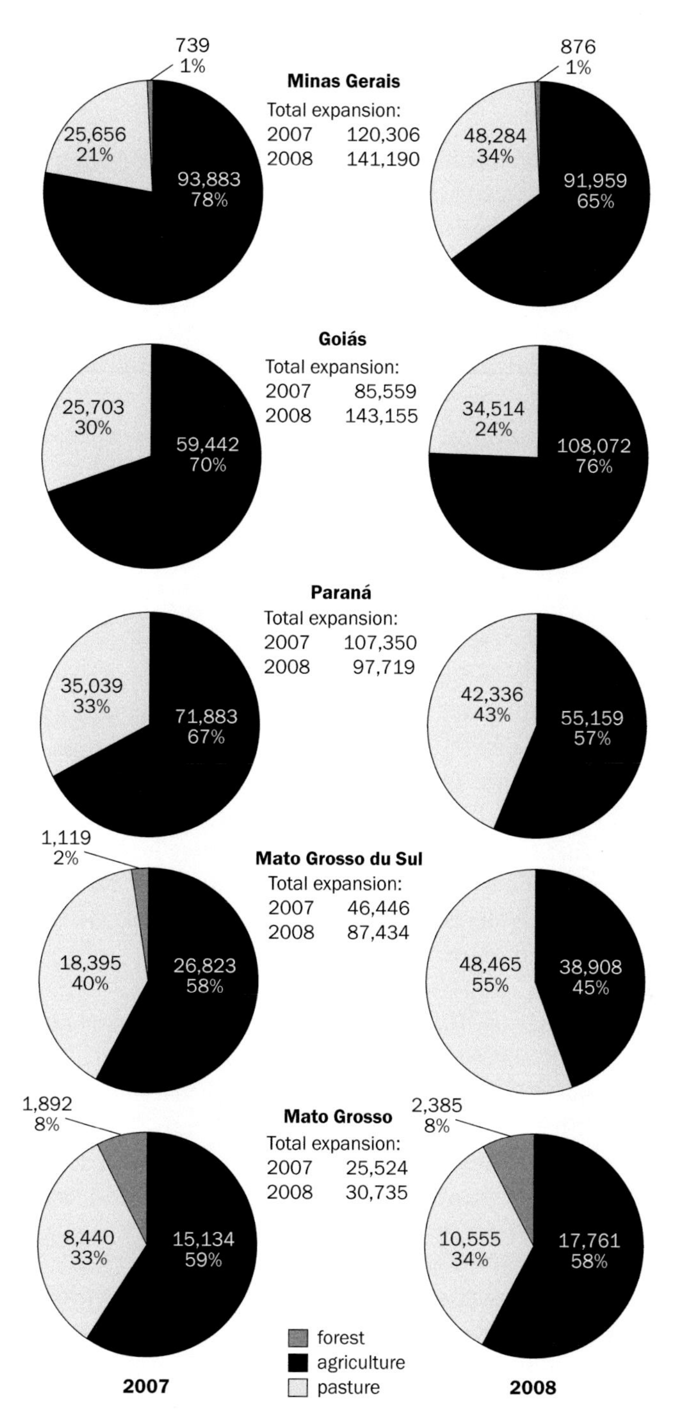

Figure 5. Area in hectares and percentages of the land use classes converted to sugarcane in 2007 and 2008 in the states of Minas Gerais, Goiás, Paraná, Mato Grosso do Sul and Mato Grosso.

Mato Grosso do Sul in 2008 when the *Pasture* and *Agriculture* classes took place over 55.4 and 44.5% of the sugarcane expansion, respectively.

For the analyzed states in this section it was verified that the *Agriculture* class decreased its contribution of displaced area for sugarcane expansion from 69.3% in 2007 to 62.2% in 2008. Conversely the *Pasture* class increased its contribution from 29.4% in 2007 to 37% in 2008 suggesting a trend of increasing sugarcane expansion over the *Pasture* class.

In Minas Gerais, Goiás and Paraná more than 99% of the sugarcane expansion was observed over the classes of *Agriculture* and *Pasture*. In Mato Grosso do Sul, 2.41% (1,119 ha) of the area of sugarcane expansion took place over the *Forest* class in 2007; while in 2008 it was insignificant (61 ha). In the state of Mato Grosso the sugarcane expansion over the *Forest* class was 7.4% (1,892 ha) in 2007 and 7.8% (2,385 ha) in 2008.

4.1.2. Analysis for the period from 2005 to 2008 in the state of São Paulo

Figure 6 shows the results for *Pasture, Agriculture, Reforestation, Forest* and *Citrus* classes which were displaced for the expansion of the sugarcane crop harvested for the first time from 2005 to 2008 in São Paulo State. During the analyzed period a sugarcane expansion of 1,810 million ha was observed. *Pasture* (53%; 960,000 ha) and *Agriculture* (44.6%; 808,000 ha) were responsible for 97.7% (1,768 million ha) of the change. About 2% (36,900 ha) of sugarcane expansion took place over the *Citrus* class and 0.31% (5,500 ha) over the *Reforestation* and *Forest* classes together. Based on the data shown in Figure 6 it is not possible to conclude that the *Pasture* class tends to increase its contribution in displaced area for sugarcane expansion in relation to the *Agriculture* class. Nevertheless in 2008 the *Pasture* class contribution is the largest (56.1%) and the *Agriculture* class is the smallest (40.6%) in the four analyzed years.

4.1.3. Analysis for the years of 2007 and 2008 in the South-Central region

Figure 7 shows the results that refer to the classes of land use that were displaced for sugarcane expansion in the most relevant producing states of the South-Central region in 2007 and 2008. In both years, *Pasture* and *Agriculture* classes were together responsible for 98.1% of the total area displaced for sugarcane expansion (2,184 thousand ha). The *Pasture* class was responsible for 45.4% (0.991 million ha) and the *Agriculture* class was responsible for 52.7% (1,152 thousand ha) of the displaced area for sugarcane (Table 3). About 1.3% of sugarcane expansion took place over the *Citrus* class (28,916 ha) and 0.58% (12,623 ha) over the *Reforestation* and *Forest* classes together (others in Figure 7). Figure 7 shows that the *Agriculture* class was more displaced than the *Pasture* class for sugarcane expansion; however,

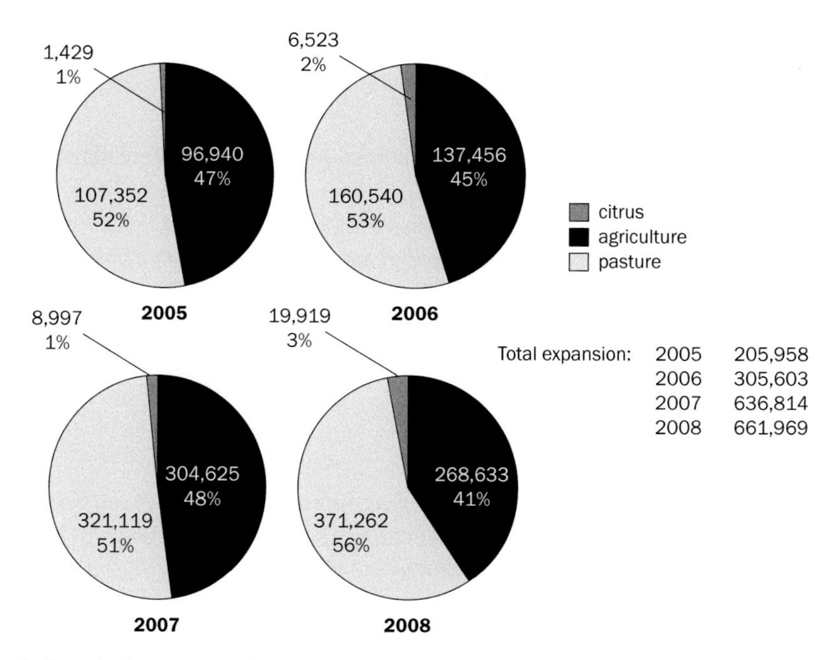

Figure 6. Area in hectares and percentages of the classes of land use that were displaced for sugarcane expansion in São Paulo State from 2005 to 2008.

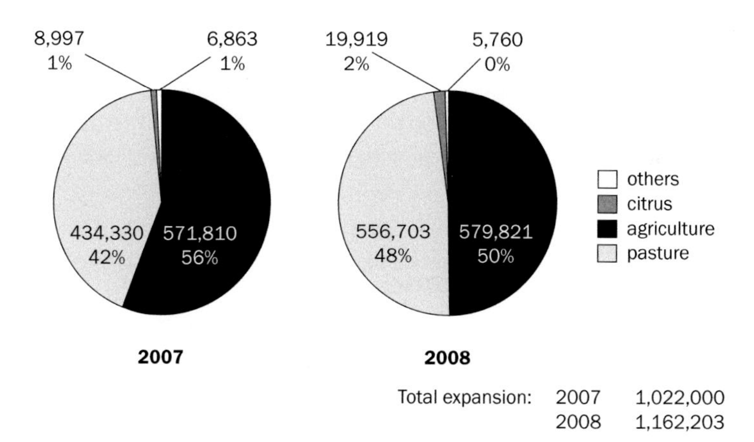

Figure 7. Area in hectares and percentages of the classes of land use displaced for the expansion of sugarcane in the most relevant sugarcane producing states in the South-Central region in 2007 and 2008.

the *Pasture* class increased from 42.5 to 47.9% its relative contribution, while the *Agriculture* class decreased its relative contribution from 55.9 to 49.9% from 2007 to 2008.[7]

Finally we can conclude that the remote sensing images obtained systematically by the Landsat and CBERS satellites enabled an accurate identification of land use classes defined in this work, which were displaced for the recent and most relevant sugarcane expansion observed in Brazil. The visual interpretation of the remote sensing images on a computer screen is hard-working, but it allows an accurate classification of the areas of interest producing a reliable thematic classification through an objective and measurable procedure.

4.2. Micro-regional secondary data

The dynamics of agricultural area and sugarcane in particular is presented in Table 4. Sugarcane expansion from 2002 to 2006 in the ten states analyzed reached 1,077 thousand ha, and more than half of this number occurred in São Paulo State. From this total expansion, 773 thousand ha displaced pasture land and 103 thousand ha displaced other crops, while only 125 thousand ha were not able to be allocated over previous productive areas. Total agricultural area growth – the sum of all crops, including sugarcane, and pastures – in the period was 3,376 thousand ha.

By looking at Table 4 it is possible to state that sugarcane expansion is relatively small comparing to total agricultural expansion in many states, particularly in Mato Grosso and Tocantins (about 1% of total agricultural expansion) and Paraná and Mato Grosso do Sul (about 16% of total agricultural expansion). Expansion of agriculture as a whole means new areas are converted to productive uses, which were former natural landscapes (forests, Cerrado savannah, natural pastureland, and so on) or idle areas. In those states where total agricultural area decreased over the analyzed period (Minas Gerais, Goiás, Maranhão and Piauí), there is no clear evidence that sugarcane expansion had taken place over non anthropized areas.

It is also important to state the significant proportion of sugarcane expansion over pasture, which represents 72% of total sugarcane expansion. By state level, also 72% of sugarcane increased over pasture in São Paulo, while in Minas Gerais, Paraná and Goiás these numbers were 51, 63, and 90%, respectively. This result is quite different from the one presented in Section 4.1 and a possible reason to explain this distortion is presented in Section 4.3.

[7] A recent study published by CONAB (2008), stated that the sugarcane expansion in the 2007 harvest in Brazil, totaling 653,722 ha, has occurred mainly over pastureland, 64.7%, followed by maize and soybean, 21.8%. New areas were responsible for 2.4% of area used for sugarcane expansion. These data were collected through 343 interviews with sugarcane mill managers.

Table 4. Area of crops and pasture displaced for sugarcane expansion by State, from 2002 to 2006 (1,000 hectares).

State	Total agricultural expansion	Sugarcane expansion			
		Total growth	Over pasture	Over other crops	N.A.[1]
São Paulo	146	639	460	115	65
Minas Gerais	-1,251	160	157	2	1
Paraná	535	92	58	2	32
Goiás	-775	63	56	0.1	6
Mato Grosso do Sul	281	45	44	0.6	0.1
Mato Grosso	4,945	31	18	3	10
Bahia	124	27	19	5	3
Maranhão	-655	16	12	0	5
Piauí	-122	3	0.2	0	2
Tocantins	148	1	0.5	0	0.6
Total	3,376	1,077	773	103	125

[1] N.A. means not allocated over previous productive area.

It is important to note that the shift-share methodology is not sufficiently robust to explain the re-allocation of land in regions that are subjected to expansions in all categories of products. In those regions, new land has been incorporated into agricultural production, which might be attributed to the conversion of forest to agriculture or to the use of previous idle areas. However, the model is not able to identify this conversion since data on deforestation are not necessarily available for the same period of time and for the same geographical unit. Even if we could assume that 100 percent of the expanded area results in natural vegetation conversion, it would not be possible to isolate the contribution of the sugarcane for this process. One possible alternative to explain the not allocated area presented above would be to assume a partial indirect effect at a micro-regional level. The indirect effect is partial because the amount of not allocated area of sugarcane would be allocated in the amount of new area proportionally of the share of sugarcane expansion to the total agriculture expansion. The assumption that sugarcane is expanding in not anthropized area, however, is not corroborated by satellite images as discussed in the previous section.

4.3. Environmental licensing reports

As a first result of this field research, it was possible to count and locate new mills projects in six different states (Table 5). Goiás is the state that has more projects under analysis, which

Table 5. Number of sugarcane mills in selected states and sampled mills information on land use change.

State	Number of mills		Mills sampled					
	Under analysis/ total [1]	Sampled	Crushing capacity (1,000 ton/ year)	Sugarcane area (1,000 ha)	Previous land use (1,000 ha) [2]			
					Pasture	Grains	Permanent crops	Natural vegetation
Goiás	53/72	4	13,000	130	37.5	32.5	no	no
Tocantins	2/4	4	7,100	91	24	12	no	10
Mato Grosso do Sul	36/47	13	40,610	494	101	yes	no	no
Mato Grosso	4/14	8	15,000	200	yes	32.5	no	0.8
Minas Gerais[3]	22/52	2	3,822	47	yes	yes	no	yes
São Paulo[3]	37/205	3	8,000	108	50.5	yes	yes	no

[1] Under analysis means the project mill is under environmental analysis.

[2] The numbers refer to those studies which presented land use changes numbers, while 'yes/no' refer to those studies which did not quantified land use.

[3] Minas Gerais and São Paulo total mills are preliminary data.

means sugarcane is expanding significantly there. The projects under analysis correspond to more than 70% of total projects, while in São Paulo, traditional producer state, it represents about 18%.

Mato Grosso do Sul is another state where new mills are majority, and where sugarcane expansion is very recent. Although the state production represents less than 3% of sugarcane production in 2007, this share is expected to increase considerably over next years. Thirty six new projects are under environmental analysis, while just eleven are already in operation, and from the thirteen new projects' Environmental Studies analyzed no forest is expected to be suppressed. All new projects needs to present EIA-RIMA and public hearings take place in the local town, where the environmental agency team helps to divulgate and motivate community to participate.

Minas Gerais is currently the third biggest sugarcane producer right after Paraná State and has a significant number of new mills that are being projected. The regions concentrating the majority of new projects are mainly in the south and south-eastern – the later including a traditional cattle raising region denominated Triângulo Mineiro. In spite of the fact that sugar mills need environmental licenses, the state law doesn't require EIA-RIMA, which means that simpler studies are made in order to get the license. These studies don't include land use changes impacts and public hearings are not necessary.

In Mato Grosso State, expansion dynamics is different. During the 80's, some sugarcane mills were implemented in order to produce ethanol, as part of PROALCOOL Plan. After decline in governmental supports and the crisis in the sector, some mills have been closed down while others have started to produce sugar as well, and thus have kept business profitable. Recently, two mills that used to be no longer in operation are requesting licenses to come back to business. Nevertheless, sugarcane production is relatively small in this state, and only two new mill projects are under analysis.

The State of Tocantins is not under important sugarcane expansion now, but it is expected to be in near future if demand for land for this crop keeps increasing. The state Environmental Agency estimates that after the implementation of North-South railway the regional transportation problem will be solved and about 50 sugar mills will be constructed. Thus, the state government wants to get ready for this huge demand over environmental licensing and planning for this land use change.

São Paulo is the most important Brazilian state for sugarcane production and the most industrialized, urbanized and occupied state. Many of existing mills are requesting licenses to expand crushing capacity and some new mills projects are also under licenses analysis. Some studies analyze the impact of this recent sugarcane expansion over other crops in this state, resulting especially in reduction of pastures, citrus and maize areas (Coelho *et al.*, 2007). Camargo *et al.* (2008) state that land rental for sugarcane has increased by an average of 12.6% from 2001 to 2006 in São Paulo, contributing to diminish expansion in São Paulo, which in turn can boost sugarcane expansion in other states.

Two significant facts regarded to land use change caused by sugarcane were verified during this field research. First, many of the projects which were considering to use pasture to cultivate sugarcane mentioned the necessity of one or two years cultivating other crops, such as soybean prior to the planting of sugarcane, in order to improve the soil quality of the low productive pasture land (regarding structure and/or fertility). This fact may partially explain the relative big proportion of sugarcane expansion over the *Agriculture* class detected in the remote sensing image analyses (Section 4.1). Since data presented in this paper are not able to verify this fact, deeper analysis are necessary to evaluate whether these crops were cultivated just to prepare the soil under pasture for sugarcane, meaning that the *Agriculture* class could possibly be overestimated (and therefore the *Pasture* class underestimated).

The other significant fact is the common use of crop rotation during sugarcane renovation process. After a number of harvests – generally five to six – sugarcane yield decreases and, therefore, the sugarcane field should be renovated. This is usually performed with an '18 months' sugarcane plant. In this case an annual food crop such as soybeans can be cultivated during the summer season. Potentially, this means that about 15 to 20% of the cultivated sugarcane area can be cultivated with an annual crop in order to improve soil quality, prevent soil erosion and contribute to food production. Although this practice is not used in all sugarcane fields, it has been disseminating fast and it is likely to be used in the majority of areas all over Brazil.

4.4. Expected sugarcane expansion and implied land use changes

Projections developed for this study are indicating that harvested sugarcane area in Brazil will reach 11.7 million ha in 2018, departing from 7.8 million ha in 2008. Area allocated for crops (soybean, maize, cotton, rice and dry beans) is expected to grow from 37.8 million ha to 43.8 million ha. Pasture area will move to the opposite direction, being reduced from 165 to 162 million ha.

Projections on regional level are presented in the Figure 8. The figure shows that the South-Central region, comprising the regions South, Southeast and Center-west Cerrados, will continue to be the most relevant and dynamic. North-northeast region is also very important but, as described in the methodology section, it is not a dynamic region in terms of growth.

Table 6 summarizes the expected growth in the South-Central region. Results show that the expansion of grains and sugarcane are fully compensated by the reduction on pasture area. Projections also confirm that cattle production is improving in terms of productivity given that the herd is increasing despite of the reduction on pasture area.

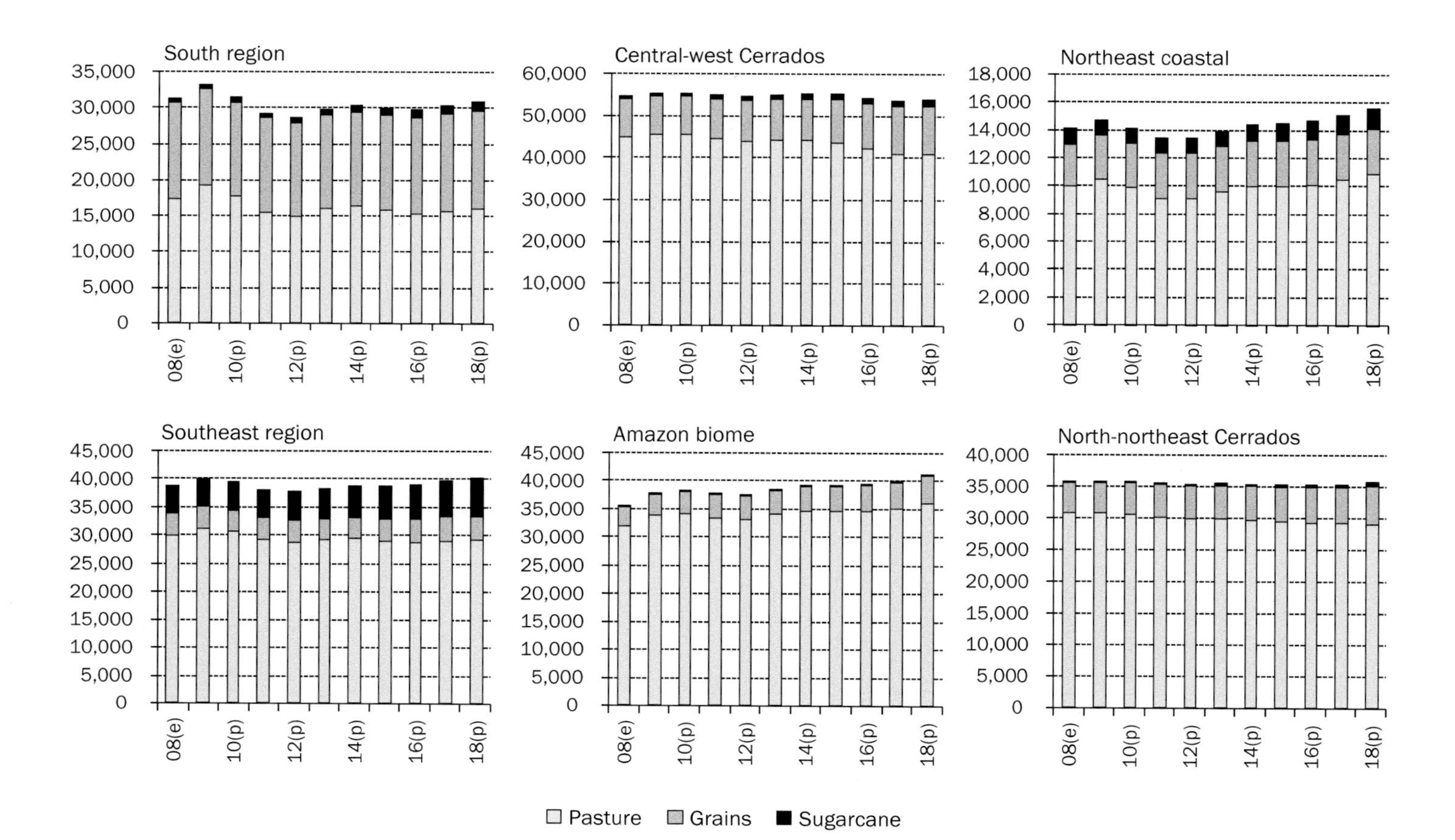

Figure 8. Projected evolution of sugarcane, grains and pasture area in agricultural regions (1,000 ha).

Table 6. South-Centre: expected land allocation for sugarcane, grains and pastures (1,000 ha and heads).

	2008	2018	Net growth
Sugarcane (ha)	6,359	9,654	3,295
Grains (ha)	26,332	29,529	3,198
Pasture (ha)	92,328	86,215	-6,113
Total (ha)	125,018	125,398	380
Cattle herd (hd)	119,399	125,501	6,102

Detailed analysis on sugarcane, crops and pasture expansion was developed for the South-Central region, in order to standardize the producing regions with the past land use change presented in the previous sections. Figure 9 shows the results obtained for São Paulo, Minas Gerais, Paraná, Goiás, Mato Grosso do Sul and Mato Grosso. Sugarcane expansion will follow trends in terms of land use change similar to the ones observed in the past.

The absolute variation shows that expansion will be larger in São Paulo, with 1.9 million ha expansion. However, the relative variation shows that Minas Gerais (98 percent growth in comparison to 2008), Paraná (98 percent expansion), Goiás (118 percent expansion) and Mato Grosso do Sul (105 percent) are the states where sugarcane will present the most dynamic expansion.

Apart from Minas Gerais, pasture is losing area in all states, while crops area is decreasing only in São Paulo. Due to this strong reduction in pasture and simultaneous growth in sugarcane and crops, the results indicate that both categories are displacing pasture. Micro-regional data in states such as Paraná and Goiás allow us to conclude that crops area displaced by sugarcane is partially compensated over pastures areas. The reduction on pastures area, as already shown in Table 6, is expected to be compensated by yields improvement. Even in states where cattle herd tend to fall, such as São Paulo and Paraná, pasture area reduction does not compromise beef and dairy production. In the Centre-West states, Goiás, Mato Grosso do Sul and Mato Grosso, pastures areas are declining and projected cattle herd is increasing, showing strong productivity gains.

It is also important to say that pasture area is expected to increase only in the Amazon Biome region. However, this expansion is taken place independently of the other regions, because cattle herd is increasing in the South-Central region, which is the sugarcane expansion region. Projections also confirm results observed by satellite images: as soon as the sugarcane increases its expansion, more pasture is displaced in comparison to grains. Pasture land displacement is majority in Minas Gerais and Centre-West states. In Paraná sugarcane expansion will push grains production to pastures area. We probably will see the same

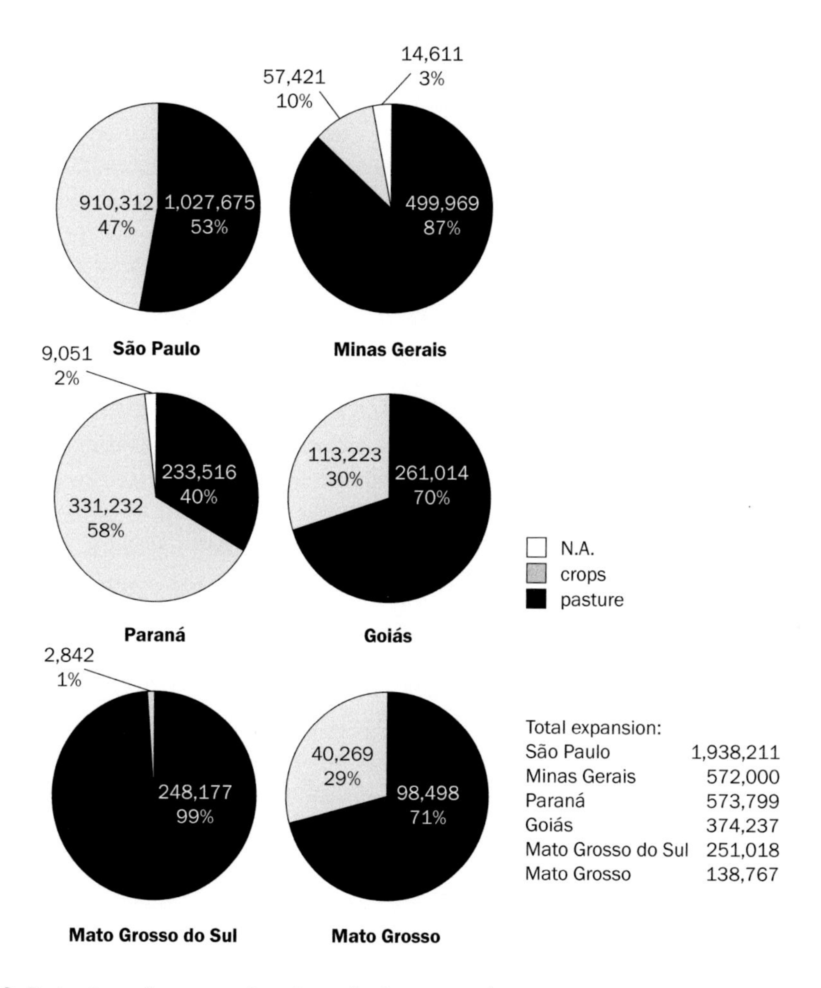

Figure 9. Projection of crops and pasture displacement due to sugarcane expansion, from 2008 to 2018 in selected Brazilian states (in ha).
Note: N.A. means not allocated over previous productive area.

process in São Paulo, although not only pasture but also grains are strongly releasing land for sugarcane.

Grains are releasing 1.6 million ha for sugarcane but, as can be observed in the Table 6, total area is expanding by 3.9 million ha, more than compensating the losses for sugarcane. Pasture area will be reduced by 6.1 million ha and sugarcane is contributing with 2.4 million ha of this total.

4.5. Options for approaching indirect effect

Indirect Land Use Change analysis is still under development, both in terms of proper definitions and methods to measure it. Regarding to its definitions, Gnansounou *et al.* (2008) classify four different sources of ILUC: spatial ILUC (displacement of prior production to other location); temporal ILUC (shifting land-use in the same location); use ILUC (shifting biomass use in the same location); and displaced activity/use ILUC (avoiding national land-use change by shifting previous activity to other country).

According to the authors, ILUC is market driven, global effect, spatial dependent and time dependent. The first two items must be analyzed in a global scale, while the last two need down-scaling analysis. Considering local analysis, although there is no consensual method in specialized literature, it is possible to use available land use data to clarify some points.

Notwithstanding that the sources of ILUC can be formally defined from a theoretical perspective, the difficulties associated to isolate and to separate the contribution of each source to the indirect effect make empirical analysis much less promising than what it appears to be. Thus, even though it is reasonable to state that the displacement of one activity as a result of the expansion of other activity may lead to an indirect land use change, as argued by Searchinger *et al.* (2008), such as deforestation, this incorporation of additional land may be happening despite of the expansion of biofuel's feedstocks production. Additionally, when the expansion of biofuel's feedstocks is taking place in conjunction with the expansion of agricultural products for food production, it is hard to prove effect-cause relations between biofuel's expansion and deforestation. This is exactly one of the fragilities of Searchinger's paper.

It is beyond of the objectives of this paper to address ILUC in a worldwide perspective. However, due to the fact that sugarcane is expanding in Brazil, it is necessary to search for arguments and data supporting the idea that sugarcane expansion is leading to an increase in the land productivity, rather than promoting incorporation of new land for food production, as grains and pasture land are displaced. Both projections and observed data give indication that this process is taking place in Brazil. As presented in Tables 6 and 7, the strong increase in pasture productivity, measured by the stocking ratio, make the Brazilian case a strong example of how hard it is to empirically prove the ILUC effect associated to the expansion of sugarcane.

As discussed in previous sections, sugarcane expansion is taking place in anthropized areas. Although this paper has no evidences regarding deforestation in Brazil, it is well known that deforestation is observed in the agricultural frontier. Brazil has two most important frontiers: the Amazon Biome region, where the Amazon Forest is located, and the North-northeast Cerrados region (also called as MAPITO region), where the larger stock of savannah land is located. Both past data and projections have shown that sugarcane is not significantly expanding in these regions. Moreover, the expansion of pastures and grains area in the Amazon Biome region, which one would argue that it is happening due to the indirect effect, are lower than the area displaced by sugarcane in the South-Central region. As presented in the Table 6, we project that pasture will lose 6 million ha for sugarcane and grains in the South-Central region for the period of 2008 to 2018. Projections for the Amazon Biome region indicate that pasture area will increase by 4 million ha for the same period (Figure 8). Those numbers show that one unity of pasture land lost does not have to be fully compensated in the frontier because productivity of the cattle production is increasing. Results of this study, therefore, support the idea that both pasture land improvement and increasing stocking rate can more than compensate land released for sugarcane and even for other crops. Regional herd data presented in Table 7 reinforces even more this evidence.

In the states where sugarcane area increased from 2002 to 2006, other crops area have also increased (exception for São Paulo), which means there is no clear reason to state that sugarcane has displaced crops which in turn could occupy natural vegetation (Table 7). A similar rational can be made for pasture land, but now including yield improvement. The

Table 7. Net growth of sugarcane, other crops, pasture land, total used area, and cattle herd from 2002 to 2006 in selected Brazilian states (1,000 ha and heads).

State	**Net growth 2002-06**				
	Sugarcane (ha)	**Other crops (ha)**	**Pasture land (ha)**	**Total used area (ha)**	**Cattle herd (heads)**
São Paulo	622	-224	-882	-484	-909
Minas Gerais	153	390	-625	-82	1,644
Paraná	74	850	-636	287	-284
Mato Grosso do Sul	41	734	-985	-210	558
Goiás	34	576	-2,041	-1,431	545
Bahia	27	492	143	661	912
Mato Grosso	25	1,634	-1,437	222	3,881
Maranhão	16	298	-463	-148	1,835
Pará	3	115	2,502	2,620	5,311
Piauí	3	206	-112	97	34
Rondônia	1	124	-364	-239	3,444
Tocantins	0.9	238	-595	-355	778
Acre	0.7	13	109	123	635
Total	1,000	5,446	-5,385	1,061	18,383

Source: PAM/IBGE, Agricultural Census/IBGE and PPM/IBGE.

states that have lost pasture land have also increased cattle herd (exception for São Paulo and Paraná), meaning there was an improvement in the cattle sector. Therefore, it is important to state that biofuels produced from biomass grown on unused arable land or resulting from yield improvements (as much of the pasture land displaced for sugarcane) have no indirect effects according to the Roundtable on Sustainable Biofuels (2008).

Thus, yields improvements in crops can be considered as area's release, meaning that the same amount of cereals, for example, is produced on a smaller area, leaving area available for other uses. *Ceteris paribus*, i.e. not considering other variables such as increase in demands, yield improvements alleviate area for other purposes. Sugarcane cultivated over these areas does not compete with land and has no indirect effects. For a total of about 1,390 thousand ha of agricultural area displaced for sugarcane verified by satellite images in the six states analyzed, 572 thousands ha were released by crops yields improvements (Table 8).

Grains, cereals and oilseeds area displaced, besides yields improvements, necessarily have to be compensated in a non-sugarcane area, although food production would be compromised. However, crops re-allocation could also take place in pastures areas, being partially compensated by cattle yield improvements. Moreover, if the expansion of food

Table 8: Agricultural area displaced by sugarcane and agricultural area compensated by yield improvement from 2005 to 2008 (São Paulo) and 2007 to 2008 for the other States.

State	Agricultural area displaced by sugarcane[1] (1,000 ha)	Average annual yield growth[2]	Agricultural area compensated by yield improvement[3] (1,000 ha)
Minas Gerais	186	4.4%	118
Goiás	168	1.9%	46
Paraná	127	4.8%	231
Mato Grosso do Sul	66	5.8%	30
Mato Grosso	33	9.5%	44
São Paulo	808	3.4%	103
Total	1,387		572

[1] Source: Figures 5 and 6.
[2] Averages are calculated from CONAB data for soybeans, maize, rice, dry beans and cotton, for the period of 1991 to 2008 (www.conab.gov.br).
[3] The baseline agricultural area (2005 for São Paulo and 2007 for the remaining states) for the selected crops were discounted by the area 'saved' due to those yield growth.

crops is larger than the amount of land displaced by sugarcane, indirect effect cannot be quantified. This is also the situation taking place in Brazil.

For global scale considerations for ILUC, however, the analysis is much more complex. In this case, it is even more difficult to determine the proper scale pertinent, considering that many countries produce and trade different products that can be related to ILUC analysis. Regarding to shifting biomass use, Brazilian sugar production has also increased from 2002 to 2006, from 22.5 to 29.6 million tons, according to UNICA (2008). Meats and all grains production have also increased significantly in the period and in the last years. Therefore, there was no need to convert land in other countries due to increase in ethanol production in Brazil. Nevertheless, all ILUC considerations can be considered preliminary and subject to improvements since the topic has been developing fast recently.

5. Conclusions and recommendations

This work was an effort to analyze land use changes due to sugarcane expansion in Brazil, contributing to the global debate about social and especially environmental benefits of ethanol. A careful look of the distribution of sugarcane shows that the crop is located and is expanding in regions that are devoted to agricultural production since a long time. Projections indicated that sugarcane expansion will continue to take place on these areas. This means that there is no sugarcane expansion in the agricultural frontier, which is the place where agricultural production has been converting natural landscapes. Thus, results are indicating that sugarcane is not directly pressing natural vegetation in any region in Brazil.

The use of different methods gives consistency to the analysis, since each one has its weakness and strengths. Remote sensing images can be considered the most reliable source of information, however they focused only in areas where sugarcane expansion took place, neglecting dynamics of other crops and pasture land. Secondary data from IBGE cover all significant productive land uses, nevertheless the data is subject to accuracy problems, especially those relating to areas dedicated to pasture. Case studies through environmental licensing reports can offer profound analysis to understand the dynamics of the mill, although they are limited in scope of coverage. The land use model projects future trends of land substitution among crops, based on past trends, but it relies on many economic assumptions.

Both remote sensing and secondary data analysis have generated similar results regarding direct land use changes promoted by sugarcane expansion. Although results are different in terms of crops and pasture land displacement, they both corroborate that sugarcane expansion has taken place with no direct effect on natural vegetation land. Furthermore, pasture is increasing its participation on the area displaced by sugarcane, and this pattern is expected to continue or even become more relevant in the future.

It is important to contextualize the LUC caused by sugarcane within the entire Brazilian debate regarding land use and occupation and other factors correlated, including agricultural and environmental public policies, international commodity markets and technology development. ILUC issues, in turn, are even more complex and correlated to many other factors. Sugarcane is concentrated in the most densely occupied state, São Paulo, which has been presented signals of saturation many years ago. In a lack of clear land use and occupation policy, sugarcane, as well as many other agricultural activities, has been expanding around close states according to local conditions, both agronomic and economic. This movement presses land valorization and thus contributes to improvements on agricultural yields (crops and pasture), as it is happening in almost all regions in Brazil.

This study concludes that the expansion of crops, except sugarcane, and pasture land is taking place despite of the sugarcane expansion. This is important because it reinforces that, even recognizing that sugarcane expansion contributes to the displacement of other crops and pasture, there is no evidence that deforestation caused by indirect land use effect is a consequence of sugarcane expansion. Results on past data and projections show that increasing cattle herd stocking rate is able to offset pasture land reduction in regions where competition for land is taking place. Increasing productivity on cattle production, therefore, also reinforces that the expansion of pasture land on the Amazon Biome is not directly promoted by the expansion of crops and sugarcane in the non-frontier regions.

It is strongly recommended that the analysis here presented continues on a regular base in order to guarantee that sugarcane activity continues to respect natural landscapes. As any other agricultural product, sugarcane also contributes to land use changes. However, as discussed here, these changes do not undermine sugarcane's environmental benefits as a renewable agricultural-based biofuel.

References

Camargo, A.M.M.P., D.V. Caser, F.P. Camargo, M.P.A. Olivette, R.C.C. Sachs and S.A. Torquato, 2008. Dinâmica e tendência da cana-de-acucar sobre as demais atividades agropecuárias, Estado de São Paulo, 2001-2006. Informações Econômicas 38: 47-66.

Coelho, S.T., P.M. Guardabassi, B.A. Lora, M.B.C.A. Monteiro and R. Gorren, 2007. A Sustentabilidade da expansão da cultura canavieira. Centro de Referencia em Biomassa – CENBIO – USP. Cadernos Técnicos da Associação Nacional de Transportes Públicos, v. 6.

Companhia Nacional de Abastecimento (CONAB), 2008. Perfil do Setor de Açúcar e Álcool no Brasil. Brasília: CONAB.

Food and Agricultural Policy Research Institute (FAPRI), 2008. FAPRI 2008: U.S. and World Agricultural Outlook. FAPRI Staff Report 08-FSR 1. Available at: www.fapri.org (accessed in October 4th, 2008).

Gnansounou, E., L. Panichelli, A. Dauriat and J.D. Villegas, 2008. Accounting for indirect land-use changes in GHG balances of biofuels: review of current approaches. École Polytechnique Fédérale de Lausanne, Working Paper 437.101, March 2008.

Instituto Brasileiro de Geografia e Estatística (IBGE), 2008b. Censo Agropecuário de 2008. Available at: www.sidra.ibge.gov.br (accessed in July 10th, 2008).

Instituto Brasileiro de Geografia e Estatística, (IBGE), 2008a. Levantamento Sistemático da Produção Agrícola. Available at: www.sidra.ibge.gov.br (accessed in July 10th, 2008).

Oliveira, A.A., M.F.M. Gomes, J. Dos Santos, L. Rufino, A.G. Da Silva Júnior, and S.T. Gomes, 2008. Estrutura e dinâmica da cafeicultura em Minas Gerais. Revista de Economia Agrícola 34: 121-142.

Roundtable on Sustainable Biofuels, 2008. Background paper (teleconference June 3rd, 2008) of the Expert advisory group and working group on GHG. Available at: http://cgse.epfl.ch/page65660.html (accessed in August 2nd, 2008).

Rudorff, B.F.T and L.M. Sugawara, 2007. Mapeamento da cana-de-açúcar na Região Centro-Sul via imagens de satélites. Informe Agropecuário. Geotecnologias, Belo Horizonte, 241: 79-86.

Searchinger, T., R. Heimlich, R.A. Houghton, F. Dong, A., Elobeid, J. Fabiosa, S., Tokgoz, D., Hayes and T. Yu, 2008. Land clearing and the biofuels carbon debt. Sciencexpress. Feb., 2008.

Sugawara, L.M.S., B.F.T. Rudorff, R.M.S.P., Vieira, A.G. Afonso, T.L.I.N. Aulicino, M.A. Moreira, V. Duarte and W.F. Silva, 2008. Imagens de satélites na estimativa de área plantada com cana na safra 2005/2006 – Região Centro-Sul. São José dos Campos: INPE (15254-RPQ/815). 74 p.

União da Indústria de Cana-de-Açucar (ÚNICA), 2008. Estatísticas de produção de açúcar no Brasil. Available at: www.unica.com.br (accessed in July 29th, 2008).

Chapter 4
Mitigation of GHG emissions using sugarcane bioethanol

Isaias C. Macedo and Joaquim E.A. Seabra

1. Introduction

The implementation of the Brazilian sugarcane ethanol program always included a continuous assessment of its sustainability. The possibilities for increasing production in the next years must consider the exciting promises of new technologies (that may lead to 50% more commercial energy/ha, from sugarcane) as well as environmental restrictions. The greenhouse gases emissions associated with the expansion are analyzed in the next sections.

2. Ethanol production in 2006 and two Scenarios for 2020

After the initial growth with the Pro-Álcool program (~12 M m^3, from 1975 to 1984) ethanol production in Brazil stabilized at this level until 2002, when the implementation of the Flex Fuel cars led to a new period of strong growth (from 12.5 M m^3 in 2002 to ~24 M m^3 in 2008; internal demand scenarios point to 40 M m^3 in 2020, with exportation in the 10-15 M m^3 range) (Carvalho, 2007; CEPEA, 2007; MAPA, 2007; EPE, 2007). Environmental legislation phasing out sugarcane burning practices, the internal demand for electricity and the opportunity with the large number of new sugarcane mills (Carvalho, 2007) are leading to a fast transition from the 'energy self-sufficient' industrial unit to a much better use of cane biomass (bagasse and trash), turning the sugarcane industry into an important electricity supplier.

The evaluation of the GHG emissions (and mitigation) from the sector in the last years (2002-2008) and the expected changes in the expansion from 2008 to 2020 must consider technology (the continuous evolution and selected more radical changes), both in cane production as in cane processing. Two (alternative) technology paths were selected:

- The *Electricity Scenario* follows the technology trends today, with commercially available technologies: the use of trash (40% recovery) and surplus bagasse (35%) to produce surplus electricity in conventional high pressure co-generation systems (Seabra, 2008).
- The *Ethanol Scenario* considers advanced ethanol production with the hydrolysis of lignocellulosic cane residues; ethanol would be produced from sucrose but also in an annexed plant with the surpluses of bagasse and of the 40% trash recovered (Seabra, 2008). This condition would lead to a smaller area (29% smaller, for the same ethanol production) than the *Electricity Scenario*; technologies may be commercial in the next ten years.

The 2006 results are based on 2005/2006 average conditions, with the best available and comprehensive data for the Brazilian Center-South Region (Macedo *et al.*, 2008). Note that GHG emissions/mitigation are evaluated for each Scenario specific conditions; Scenario implementation schedules are not presented (or needed) for the objective of this study. However, it must be said that the *Electricity Scenario* implementation is occurring now in all Greenfield operations, and already in some retrofit of existing units. The *Ethanol Scenario* as proposed still depends on technological development of the biomass hydrolysis/fermentation processes, and it would take longer to be implemented to a significant level in the context of the Brazilian ethanol production (Seabra, 2008).

The essential parameters for 2006 and the two 2020 Scenarios are presented in Table 1 (Cane production) and Table 2 (Cane processing). The data used for 2006 is for a sample of 44 mills (100 M t cane/season), all in the Brazilian Center South. Data have been collected/processed for the last 15 years, for agriculture and industry, for the CTC 'mutual benchmarking'.

The ethanol transportation (sugar mill to gas station) energy needs are (Seabra, 2008):
2006: 100% road (trucks), 340 km (average), 0.024 l diesel/(m^3.km) (energy consumption).
2020: 80% road, with average transport distance and diesel consumption as in 2006; 20% pipeline, 1000 km (average), 130 kJ/(m^3.km) (pipeline energy consumption).

The hydrolysis/fermentation parameters in the *Ethanol Scenario* correspond to a SSCF process, expected to be commercial before 2020, as seen in Table 3.

3. Energy flows and lifecycle GHG emissions/mitigation

The systems boundaries considered for the energy flows and GHG emissions and mitigation include the sugarcane production, cane transportation to the industrial conversion unit, the industrial unit, ethanol transportation to the gas station, and the vehicle engine (performance). Methodologies use data and experimental coefficients as indicated in the tables, and in some cases IPCC (IPCC, 2006) defaults; details are presented in Macedo *et al.* (2008), Seabra (2008) and Macedo (2008). The CO_2 (and other GHG) related fluxes are:

- CO_2 absorption (photosynthesis) in sugarcane; its release in trash and bagasse burning, residues, sugar fermentation and ethanol end use. These fluxes are not directly measured (not needed for the net GHG emissions).
- CO_2 emissions from fuel use in agriculture and industry (including input materials); in ethanol transportation; and in equipment/buildings production and maintenance.
- Other GHG fluxes (N_2O and methane): trash burning, N_2O soil emissions from N-fertilizer and residues (including stillage, filter cake, trash)
- GHG emissions mitigation: ethanol and surplus bagasse (or surplus electricity) substitution for gasoline, fuel oil or conventional electricity.

Table 1. Basic data: sugarcane production.

Item	Units	2006 [a]	2020 scenarios [b]
Sucrose content	% cane stalks	14.22	15.25 [c]
Fiber content	% cane stalks	12.73	13.73 [d]
Trash (db) [e]	% cane stalks	14	14
Cane productivity	t cane/ha	87.1	95.0
Fertilizer utilization [f]			
P_2O_5	kg/(ha.year)	25	32
K_2O	kg/(ha.year)	37	32
Nitrogen	kg/(ha.year)	60	50
Lime [g]	t/ha	1.9	2.0
Herbicide [h]	kg/ha	2.2	2.2
Insecticide [h]	kg/ha	0.16	0.16
Filter cake application	t (db)/ha (% area) [i]	5 (70%)	5 (70%)
Stillage application	m^3/ha (% area) [j,k]	140 (77%)	140 (77%) [l]
Mechanical harvesting	% area	50	100 [m]
Unburned cane harvesting	% area	31	100 [m]
Diesel consumption	L/ha	230	314

[a] CTC's database (44 mills in Center-South of Brazil, equivalent to ~100 Mt cane/year) (CTC, 2006a).

[b] Author's projections; Scenarios are Electricity and Ethanol.

[c] 2020: increasing 1 point (%) in 15 years (variety development and better allocation).

[d] Apparent fiber increasing with increase in green cane harvesting (trash).

[e] Hassuani *et al.* (2005).

[f] Total averages, including: fertilizer use in plant and ratoon cane, in areas with and without stillage; full description in Macedo *et al.* (2008). For Scenario 2020 Ethanol averages are slightly lower (~4%) due to larger stillage production/utilization.

[g] Utilized essentially at planting.

[h] Macedo (2005a).

[i] Reforming areas: areas where sugarcane is re-planted, after the 6 year cycle.

[j] Ratoon areas: areas where sugarcane is cut to grow again, without re-planting

[k] It is considered that all stillage is used only in the 'ethanol cane area', but keeping the suitable level of application (~140 m^3/ha). For 2020 Ethanol scenario, see Note l.

[l] In the 2020 Ethanol scenario more stillage would be produced, from the ethanol derived from hydrolysis. Stillage application would reach larger ratoon areas.

[m] Considering the legislation and phase out schedules for cane trash burning in SãoPaulo.

Table 2. Basic data: sugarcane processing.

Item	Units	2006 [a]	2020 electricity [b]	2020 ethanol [b]
Bagasse use		Low pressure cogeneration	Advanced cogeneration	Biochemical conversion
Electricity demand	kWh/t cane	14.0	30	[c]
Mechanical drivers	kWh/t cane	16.0	0	0
Electricity surplus	kWh/t cane	9.2 [d]	135 [e]	44 [f]
Trash recovery	% total	0	40	40
Bagasse surplus	% total	9.6	0 [g]	0 [g]
Ethanol yield	l/t cane	86.3	92.3 [h]	129

[a] CTC information (CTC, 2006b).
[b] Authors' projections.
[c] 30 kWh/t cane + 130 kWh/t hydrolyzed biomass (dry basis).
[d] Cogen's data; only 10% of the mills use higher pressure boilers, and the remaining 90% still use 21 bar/300 °C, with very low electricity surplus.
[e] All mills operating at 65 bar/480 °C, CEST systems; process steam consumption ~340 kg steam/t cane, and using recovered trash (40%).
[f] A hypothetical mill operating at 65 bar/480 °C, 'pure' cogeneration; using ~340 kg steam/t cane.
[g] All biomass (bagasse and 40% trash) is used for power generation or ethanol production.
[h] Only the increase in sucrose % cane was considered.

Table 3. Bioconversion parameters (SSCF process with dilute acid pretreatment)[a].

Hydrolysis	95 % (cellulose); 90% (hemicellulose)
Fermentation	95% (glucose); 85% (other sugars)
Energy demand	
Electricity	130 kWh/t (db)
Steam	
Pre-treatments (kg/kg db)	0.45 (13 bar); 0.25 (4.4 bar)
Distillation (kg/l et)	3.0 (2.5 bar); 0.05 (22 bar)
Concentration (kg/l et.)	0.2 (1.7 bar)

[a] Based on Aden *et al.* (2002); details in Seabra (2008).

The GHG emissions associated with direct land use change (LUC) are estimated separately in the next section, where the possible indirect impacts of land use change (ILUC) are also discussed for the specific case of the expansion of ethanol production in Brazil. The energy use/conversion for 2006 and for each 2020 Scenario is presented in Table 4.

The corresponding GHG emissions for are in Table 5. Note that the differences in total emissions are strongly dependent on the co-products credits. The large difference between 2006 and the 2020 Electricity Scenario is due to an actual increase in the system energy efficiency (much larger energy output). An analogous increase in energy output occurred between 2006 and the 2020 Ethanol Scenario, but note that the change is an increase in ethanol output (rather than in electricity) and also the emissions are presented in kg CO_2 eq/m^3 ethanol. In the 2020 Ethanol Scenario the volume of ethanol produced/unit area (or ethanol/ t cane) is 1.4 times larger than in the 2020 Electricity Scenario.

It is important to remember that the 2006 data (and results) correspond to the average values of the parameters; even for a homogeneous set of producers (Brazil Center South region) differences in processes (agricultural and industrial) impact energy flows and GHG emissions. For the sample used, the variation of main production parameters and the

Table 4. Energy balance in anhydrous ethanol production (MJ/t cane).

	2006	**2020 electricity**	**2020 ethanol**
Energy input	235	262	268
Agriculture	211	238	238
Cane production	109	142	143
Fertilizers	65	51	50
Transportation	37	45	45
Industry	24	24	31
Inputs	19	20	25
Equip./buildings	5	4	6
Energy output	2,198	3,171	3,248
Ethanol [a]	1,926	2,060	2,880
Electricity surplus [b]	96	1,111	368
Bagasse surplus [a]	176	0.0	0.0
Energy ratio	9.4	12.1	12.1

[a] Based on LHV (Low Heating Value).

[b] Considering the substitution of biomass-electricity for natural gas-electricity, generated with 40% (2006) and 50% (2020) efficiencies (LHV).

Table 5. Total emission in ethanol life cycle (kg CO_2 eq/m^3 anhydrous) [a].

	2006	2020 electricity	2020 ethanol
Cane production	416.8	326.3	232.4
Farming	107.0	117.2	90.6
Fertilizers	47.3	42.7	23.4
Cane transportation	32.4	37.0	26.4
Trash burning	83.7	0.0	0.0
Soil emissions	146.3	129.4	92.0
Ethanol production	24.9	23.7	21.6
Chemicals	21.2	20.2	18.5
Industrial facilities	3.7	3.5	3.2
Ethanol distribution	51.4	43.3	43.3
Credits			
Electricity surplus [b]	-74.2	-802.7	-190.0
Bagasse surplus [c]	-150.0	0.0	0.0
Total	268.8	-409.3	107.3

[a] Emissions for hydrous ethanol/m^3 are about 5% less than values verified for anhydrous ethanol.
[b] Considering the substitution of biomass-electricity for natural gas-electricity, generated with 40% (2006) and 50% (2020) efficiencies (LHV).
[c] Considering the substitution of biomass fuelled boilers (efficiency = 79%; LHV) for oil fuelled boilers (efficiency = 92%; LHV).

corresponding response to each single parameter variation in GHG emissions are shown in Figure 1.

Note that the electricity surplus and the bagasse surplus show very large variation now, when a few mills have started to export large amounts of electricity. The net GHG avoided emissions, including the ethanol substitution for gasoline and considering the engines performances in Brazil (based on the experience with the fleet of 23 M vehicles, in the last 30 years, with E-24, E-100 and Flex Fuel engines) is shown in Table 6.

The use of the allocation (energy) criterion for the co-products (with the whole GHG emissions associated with cane and ethanol production being distributed among ethanol, electricity and surplus bagasse according to their energy content, and with no co-product credits considered in the net emission) is compared to the use of the substitution criterion (with the mitigation derived from ethanol, electricity and surplus bagasse use being considered as well as all emissions from cane and ethanol production) in Figure 2; the substitution criterion results are detailed in Table 6.

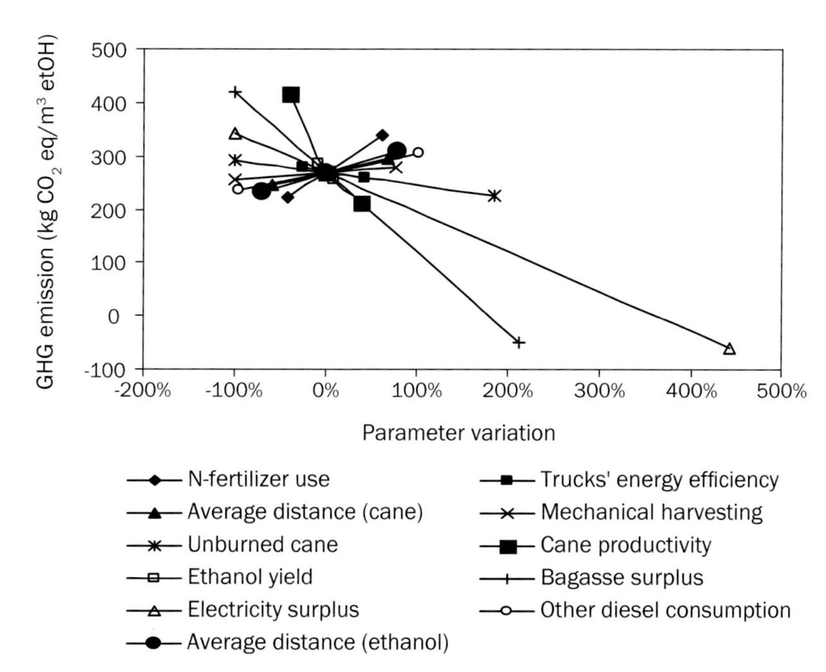

Figure 1. GHG emissions variation in response to single parameter variation; including co-product credits (2006 only).

Table 6. Avoided emissions due to ethanol use (t CO_2 eq/m^3 hydrous or anhydrous; substitution criterion for the co-products).

	Ethanol use [a]	**Avoided emission** [b]	**Net emission** [c]
2006	E100	-2.0	-1.7
	E25	-2.1	-1.8
2020 electricity	E100	-2.0	-2.4
	FFV	-1.8	-2.2
	E25	-2.1	-2.5
2020 ethanol	E100	-2.0	-1.9
	FFV	-1.8	-1.7
	E25	-2.1	-2.0

[a] E100, or HDE: hydrous ethanol in dedicated engines; FFV: hydrous ethanol in flex-fuel engines; E25: anhydrous ethanol (25% volume) and gasoline blend.

[b] Avoided emission (negative values) due to the substitution of ethanol for gasoline; fuel equivalencies verified for each application in Brazil (Macedo *et al.*, 2008).

[c] Net emission = (avoided emission due to ethanol use) + (ethanol life cycle emission). Co-products credits are included.

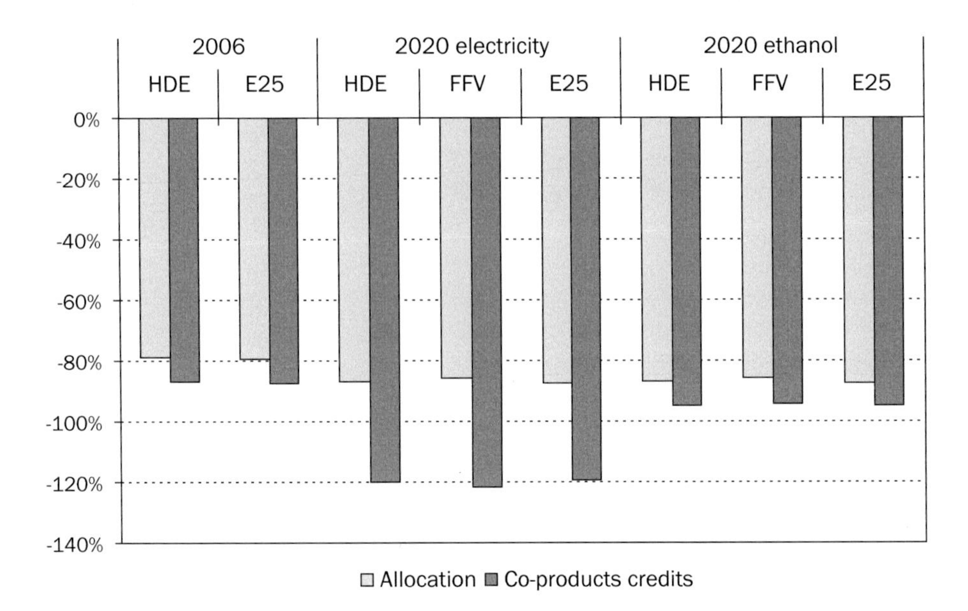

Figure 2. GHG mitigation with respect to gasoline: allocation or co-products credits.

As expected from the energy balances, the use of ethanol from sugarcane substituting for gasoline leads to a very important GHG emissions mitigation; this is due mostly to the use of cane residues as the source of energy for processing sucrose to ethanol. However, the much more efficient use of the cane residues is leading to entirely offset the gasoline emissions, and beyond that, as shown in Figure 2 for the 2020 Electricity scenario. A separate accounting of the gains with electricity and ethanol (with allocation of the emissions) still shows gains in the 2020 Scenarios, due to increase efficiencies/productivity in both cane production and processing.

4. Land use change: direct and indirect effects on GHG emissions

The variation in carbon stocks (both in soils and above ground) due to changes in the land use is included in the national carbon inventories, and evaluation methodologies have been established. The large number of parameters (culture type; soil; cultivation practices; local climate) and the lack of sufficient and adequate information for many cases lead to large error estimates for the default values, both for the basic soil type/climate carbon stocks for native vegetation and for the relative (main parameters) stock change factors (IPCC, 2006). The use of adequate local data is recommended.

Recently the so called indirect land use change (ILUC) impact in emissions is being discussed; the debate shows that we do not have suitable tools (methodologies) or sufficient data to reach acceptable, quantified conclusions about ILUC impacts on GHG emissions, globally.

However, local conditions in Brazil indicate a good possibility of significant increases in ethanol production without increasing ILUC emissions. Both LUC and ILUC impacts on emissions for ethanol are considered below.

4.1. Land use change with the ethanol production in Brazil: past and trends

Brazil has 28.3% (~440 M ha) of all original forests in the world, over a total surface of 850 M ha. From the agricultural land, only 46.6 M ha are used for grain; 199 M ha correspond to 'pasture'; large fraction of this is somewhat degraded land (extensive grazing, not planted pasture).

Sugarcane for ethanol today (2008) uses only 4 M ha, slightly more than 1% of the arable land in Brazil. Ethanol production increased fast with the Pro-Alcool Program until 1985, when it reached 11.8 M m^3; stabilized at this level, and in 2002 the production was 12.5 M m^3. *There was no land use change with cane for ethanol in the period from 1984 to 2002.*

Ethanol production growth re-started only in 2002, to an expected value of 26 M m^3 in 2008 (MAPA, 2007; CONAB, 2008). The expansion area (sugar and ethanol) from 2005 to 2008 was 2.2 M ha in the Center-South (Nassar, 2008); ethanol used 49% of the sugarcane in 2002, and 55% in 2008.

The patterns of land use change, as well as the changes in cane culture procedures determine the associated impacts on GHG emissions. For land use change, a recent analysis (Nassar et al, 2008) uses satellite images (Landsat and CBERS) available since 2003 for State of São Paulo, and 2005 for other States; and secondary data (based on IBGE data, for the whole region, from 2002 to 2006) is also analyzed for each micro-region using a Shift Share model. A comprehensive field survey was reported by CONAB (CONAB, 2008) for the LUC involving sugarcane from 2007 to 2008. Data was also obtained from the Environmental impact reports (EIA – RIMA) needed for licensing new increases in sugarcane area (Nassar et al, 2008); they refer also to the next years expected LUC.

Some of the main conclusions for the changes from 2002 to 2008 are:

1. Sugarcane always substitutes for established crops, or pasture lands; for economic reasons, and with the large availability of low productivity pasture lands associated to some pasture area conversion to higher efficiency systems, very small advances in native vegetation (forests, cerrados) areas are observed. In some cases degraded pasture lands are cultivated for one or two years with soybeans, to improve soil conditions before using for sugarcane. Intercropping (rotation of sugarcane every five or six years, before a new planting, with other crops) is becoming a widespread practice. Satellite data for the last two years (2007/08 and 2008/09) for the cane expansion areas in the six Center South cane producer states (total 2.18 M ha) indicates their origin: 53% from Agriculture; 45% from Pastures; 1.3% from Citrus plantations; and only 0.5% from Arboreal Vegetation

(native or anthropic), including wood plantations (eucalyptus or pinus). The CONAB survey (CONAB, 2008) indicates for 'new areas' (not all related to native vegetation) only 1.5% of the expansion, in 2007/08. All studies indicate that Pasture lands utilization is surpassing Agriculture land utilization (for cane) in the last years, in many areas.

2. The field survey (CONAB, 2008) indicates for 2007/08 LUC the largest Agriculture area substituted was soybean, followed by maize. The use of Pasture lands is also related to the conversion from low productivity pasture (both native and some planted pasture: degradation from inadequate management, and no fertilizers) (Macedo, 2005b), to high productivity pastures, liberating areas. Estimates indicate today 150 M ha of cultivated pasture land, and 70 M ha of 'natural' pastures.
3. Most of the expansion (94%) for sugar from 1992 to 2003 occurred around the existing sugar mills, in the Center South; now sugarcane moves to the West and North of the region, in the States of Goiás, Mato Grosso, Minas Gerais and Mato Grosso do Sul. This is the trend for the next decade.

The analysis (Nassar *et al.*, 2008) included the projected patterns of land use change for the sugarcane expansion to 2020, considering land availability, biomes and reserved areas; the response to prices/costs, demand and competition in Brazil and outside.

4.2. Soil and above ground carbon stocks

A recent study (Amaral *et al.*, 2008) on ethanol production sustainability included data on below and above ground carbon stocks for sugarcane (both burned and green cane harvesting conditions) in Brazil, as well as for the most important replaced crops and vegetation. The data was obtained from more than 80 reports in the last 8 years; a selection was made to yield comparable results (for soil types, soil depths, methodology, cultural practices).

Table 7 shows some results for soil carbon from the survey, as well as the default values calculated with the IPCC recommendations (IPCC, 2006). The IPCC based values correspond to the specific soil types (High or Low activity clay, HAC or LAC), climate, crop type and cultivation practices for each crop. The experimental data indicates the soil types (HAC, LAC or Sandy) and some cultural practices, always for 20 cm depth. Selected values were used to evaluate the soil carbon stock change with land use change, for each specific case (last column).

Table 8 (Amaral *et al.*, 2008) shows the experimental values for the sugarcane and the main replaced crops/pasture above ground carbon.

Table 7. Soil carbon content for different crops (t C/ha).

Crop	IPCC defaults [a]		Experimental [b]		Selected values
	LAC	HAC	HAC	Other	
Degraded pasturelands	33	46	41	16 [c]	41
Natural pasturelands	46	63	56		56
Cultivated pasturelands	55	76	52	24 [c]	52
Soybean cropland	31	42	53		53
Maize cropland	31	42	40		40
Cotton cropland	23	31	38		38
Cerrado	47	65	46		46
Campo Limpo	47	65	72		72
Cerradão	47	65	53		53
Burned cane	23	31	35-37	35 [d]	36
Unburned cane	60	83	44-59		51

[a] Based on IPCC parameters indicated, IPCC, 2006
[b] Amaral *et al.*, 2008 (all 0-20 cm).
[c] Sandy soils.
[d] LAC soils.

Table 8. Above ground carbon stocks (t C/ha)[a].

Degraded pasturelands	1.3
Cultivated pasturelands	6.5[b]
Soybean croplands	1.8[c]
Maize croplands	3.9
Cotton croplands	2.2[d]
Cerrado *sensu strictu*	25.5[e]
Campo Limpo	8.4[f]
Cerradão	33.5[g]
Unburned cane	17.8

[a] Amaral *et al.* (2008).
[b] LAC soils.
[c] HAC soils.
[d] General value.
[e] Areas with more than 20 years without burning.
[d] Areas with 3 years without burning.
[e] Areas with 21 years without burning.

4.3. Estimated emissions from LUC

For the changes from 2002 to 2006 (areas closer to the existing mills) soil types were frequently HAC, and some of the cane was burned; for the expansion now and in next decade, soils will be closer to LAC (and for 2020, 100% green cane harvesting is assumed). The trends for land use change until 2020 are discussed in the next sections.

It is assumed that at least 70% of the pasture land used for cane is not planted pasture, with varying degrees of degradation. Using the values in Tables 7 and 8, and the areas for each type of vegetation replaced with sugarcane, the total carbon stock change was evaluated and divided by a 20 year period. For the above ground carbon stock, only the values corresponding to perennial vegetation were considered. Results are in Table 9.

Note that in all Scenarios there is a net reduction in emissions (close to 100 kg CO_2 eq/m^3 ethanol); this was expected, since the expansion areas for sugarcane include a very small fraction of native lands with high carbon stocks, and some degraded land. The specific situation for land availability, the environmental restrictions and local economic conditions (relative crop values and implementation costs), discussed in the section *Ethanol in the*

Table 9. Emissions associated with LUC to unburned cane.

Reference crop	Carbon stock change[a] (t C/ha)	Emissions (kg CO_2 eq./m^3)		
		2006	2020 electricity	2020 ethanol
Degraded pasturelands	10	-302	-259	-185
Natural pasturelands	-5	157	134	96
Cultivated pasturelands	-1	29	25	18
Soybean cropland	-2	61	52	37
Maize cropland	11	-317	-272	-195
Cotton cropland	13	-384	-329	-236
Cerrado	-21	601	515	369
Campo Limpo	-29	859	737	527
Cerradão	-36	1,040	891	638
LUC emissions[b]		-118	-109	-78

[a] Based on measured values for below and above ground (only for perennials) carbon stocks.
[b] Considering the following LUC distribution – 2006: 50% pasturelands (70% degraded pasturelands; 30% natural pasturelands), 50% croplands (65% soybean croplands; 35% other croplands); 2020: 60% pasturelands (70% degraded pasturelands; 30% natural pasturelands); 40% croplands (65% soybean croplands; 35% other croplands). Cerrados were always less than 1%.

specific Brazilian context, indicate that LUC motivated GHG emissions will not impact ethanol production growth in Brazil in the time frame considered (2020).

It must be noted that the above ground carbon in the sugarcane plant is relatively high, and even with its annual harvesting the change from any of the other crop, or even a *campo limpo*, to sugarcane will produce an additional carbon capture (corresponding to differences in the 'average' above ground carbon in the plants). This was not included here, since it has not been considered in the IPCC methodology.

4.4. Indirect land use change effects on GHG emissions of biofuels worldwide

For most land use changes anywhere some impacts (including in GHG emissions) may happen; and in our increasingly globalized economy indirect LUC impacts may occur. However some of the hypotheses and tools leading to the initial quantification of the impacts of biofuels production (Gnansounou *et al.*, 2008), as presented today, are clearly not suitable:

- A key issue for the models is the correct description of the drivers to LUC, everywhere; but many agricultural products are interchangeable, and (increasingly) traded globally; and the drivers of LUC vary in time and regionally. 'Equilibrium' conditions are not reached. Drivers are established by *local* culture, economics, environmental conditions, land policies and development programs. The development of a range of methodologies and the acquisition/selection of suitable data are needed to reach acceptable, quantified conclusions on ILUC effects. The growing consensus over this problem is summarized in the recent letter from 28 scientists to the CARB (M.D. Nichols, personal communication): '...a severe lack of hard empirical data'... (the need to) 'further study highly controversial and speculative indirect land use changes... (for the) necessary time over the next five years... before incorporating any of these indirect impacts in (the LCFS) standard'.
  Simplifying methodologies (looking to 'regions' in the world, therefore losing the global implications; or relying on indexes for too large areas, to by-pass the lack of data; or distributing the total 'estimated' ILUC emissions equally among all biofuels) would lead to still less accurate results.
- The land used for agriculture today is ~1300 M ha, *excluding pasture lands*; biofuels use less than 1.5% of that; and possibly less than 4% in 2030 (IEA, 2006). Today's distribution of production among regions/countries has never considered GHG emissions; it was determined by the local/time dependent drivers (including subsidies and food security considerations). The better knowledge of those drivers and their effects, and its use to re-direct land use as possible over all the agricultural and pasture lands worldwide, would be much more effective than just to work on the 'marginal' biofuels growth areas. We should not simply take as 'unchangeable' the huge context of today's agriculture.
- Increases in agricultural productivity, energy end-use efficiencies and the use of other energy renewable resources in the next decades may be expected, changing energy

demand and required areas for energy production, and they can entirely change the 'future' ILUC impacts of biofuels.

4.5. ILUC effects from ethanol in the specific Brazilian context

In general, exceptions (biofuel sources with no LUC indirect GHG emissions) have been considered: waste or residues; use of marginal or degraded land; unused or fallow arable land; or improving yields in currently used land. Looking at the scenarios for ethanol production in Brazil, and the land use in Brazil today (in the context of the available land) we note that:

- Most scenarios (based on Internal Demand plus some hypotheses for Exports) indicate a total of ~ 60 M m^3 ethanol in 2020 (CEPEA, 2007; Carvalho, 2007; EPE, 2007), corresponding to 36 M m^3 more than in 2008. For the 2020 conditions, *the additional area needed will be only 4.9 M ha (Scenario Electricity) or 3.5 M ha (Scenario Ethanol).* Since the Scenario Ethanol would not be implemented (even if technically successful, and competitive) in time, we may expect ~5.1 M ha of new cane area, until 2020.
- Agricultural production (crops) uses a small fraction of the total area, and only 18.5 % of the arable land (Table 10). Pasture land (200 M ha) is nearly 60% of the arable land. Sugarcane for ethanol uses only 1% of the arable land, and the Land Available (not including the conversion of pasture lands) is twenty times larger. The new area needed for sugarcane until 2020 (5.1 M ha) is only 8% of the total crop area today, or *2.5% of the pasture area today.*
- The conversion of low quality pasture land to higher efficiency productive pasture is liberating areas for other crops. The average heads/ha in Brazil was 0.86 (1996); and 0.99 (2006), with nearly 50% planted pasture (IBGE, 2006). In the State of São Paulo the average was 1.2-1.4 in the last years. The conversion of low grade pasture could release ~30 M ha for other uses.
- Sugarcane expansion is smaller than the expansion of pasture and crops; and in the places where sugarcane expands the eventual competition products (crops and cattle) also expand. *The expansion for other agricultural crops and pasture is taking place independently of sugarcane expansion.* In the period from 2002 to 2008 the sugarcane expansion displaced Pasture and Crops (CONAB, 2008; Nassar, 2008) as follows: crop area displaced, 0.5% (but crop area increased 10%, and cereal + oilseeds production increased 40%); pasture area displaced, 0.7%; total pasture area decreased 1.7% (but beef production grew 15%).

Within its soil and climate limitations, the strict application of the environmental legislation for the new units, and the relatively small areas needed, the expansion of sugarcane until 2020 is not expected to contribute to ILUC GHG emissions.

Table 10. Land use in Brazil: selected uses (2006) (UNICA, 2008; Scolari, 2006; FAO, 2005; IBGE, 2005).

Land use	Area, M ha	% of arable land	% cultivated land
Total land	850		
Forests	410		
Arable land	340 (40%)	100.0	
Pasture land	200	58.8	
Cultivated land (all crops)	63	18.5	100.0
Soybean	22	6.5	34.9
Maize	13	3.8	20.6
Sugarcane (total)	7	2.1	11.1
Sugarcane for ethanol	3.5	1.0	5.6
Available land	77	22.6	122.2

5. Conclusions

The analyses of the GHG emissions (and mitigation) with ethanol from sugarcane in Brazil in the last years (2002-2008) and the expected changes in the expansion from 2008 to 2020 show that:

- The large energy ratios (output renewable/input fossil) may still grow from the 9.4 value (2006) to 12.1 (2020) in two Scenarios: the better use of cane biomass to generate surplus electricity (2020 Electricity Scenario: already under implementation) or to produce more ethanol (2020 Ethanol Scenario: depending on technology development). The Ethanol Scenario, if fully implemented, would reduce the area needed by 29%.
- The corresponding GHG mitigation (with respect to gasoline), for ethanol use in Brazil, would increase from the 79% (2006) to 86% (2020) if only the ethanol is considered (with emissions allocation to co-products), or from 86% (2006) to 95% or 120% (2020: Ethanol or Electricity Scenarios) if all co-products credits and emissions are considered for ethanol (substitution criterion).
- LUC due to ethanol expansion started in 2002 (ethanol production was constant at the 12 M m^3 level, since 1984). In the expansion, land availability, the environmental restrictions, the relatively small area used for expansion and the local economic conditions (relative crop values and implementation costs) led to very small use of native vegetation lands (<1%), and large use of low productivity pasture lands and some crop areas: soy and maize. LUC derived GHG emissions were actually negative in the period 2002-2008. The growth scenarios for 2020 (~reaching 60 M m^3 ethanol) indicate the need for relatively small areas (~5 M ha) as compared to the availability (non used arable lands, or even degraded pasture lands); the trend is the use of more pasture lands and less crop areas, in the expansion. Again, very little impact (if any) on LUC GHG emissions are expected.

- Suitable evaluations (even estimates) of ILUC impact in emissions are far from possible today, due to the lack of adequate methodologies and corresponding (global) data. However local conditions in Brazil indicate a good possibility of significant increases in ethanol production without increasing ILUC GHG emissions:
  - The area needed for expansion (~5 M ha, until 2020) is very small when compared with the areas liberated with increased cattle raising efficiency (30 M ha) and other non used arable lands.
  - Sugarcane expansion has been independent of (and much smaller than) the growth of other agricultural crops, in the same areas. In all sugarcane expansion areas the eventual competition products (crops and beef production) also expanded.

References

Aden, A., M. Ruth, K. Ibsen, J. Jechura, K. Neeves, J. Sheehan, B. Wallace, L. Montague, A. Slayton and J. Lukas, 2002. Lignocellulosic biomass to ethanol process design and economics utilizing co-current dilute acid prehydrolysis and enzymatic hydrolysis for maize stover. Technical report TP-510-32438, NREL, Golden, CO, USA.

Amaral, W.A.N., 2008. Environmental sustainability of sugarcane ethanol in Brazil. In: P. Zuurbier and J. van de Vooren (eds.) Sugarcane ethanol, Wageningen Academic Publishers, Wageningen, the Netherlands, pp. 113-138.

Carvalho, E.P., 2007. Etanol como Alternativa Energética. UNICA: presentation to the Casa Civil, Presidência da República, Brasília, Brazil.

CEPEA, 2007. Cenários (oferta e demanda) para o setor de cana de açúcar. CEPEA (Internal Report), Esalq, Piracicaba, Brazil.

CONAB, 2008. Acompanhamento da safra Brasileira: Cana de Açúcar. Companhia Nacional de Abastecimento, MAPA, Brasilia, Brazil.

CTC, 2006a. Controle mútuo agrícola anual - safra 2005/2006. Centro de Tecnologia Canavieira, Piracicaba, Brazil, 126 p. + anexos.

CTC, 2006b. Technical information provided by scientists and engineers to I Macedo and J Seabra. Piracicaba: Sugarcane Technology Centre, Brazil.

EPE, 2007. Plano Nacional de Energia 2030. Empresa de Planejamento Energético, MME, Brazil.

FAO, 2005. Food and Agriculture Organization, U.N. Statistical Databases, Agriculture. http://www.fao.org/faostat.

Gnansounou, E., L. Panichelli, A. Dauriat and J.D. Villegas, 2008. Accounting for indirect land-use changes in GHG balances of biofuels: review of current approaches. École Polytechnique Fédérale de Lausanne, Working Paper 437.101.

Hassuani, S.J., M.R.L.V. Leal and I.C. Macedo, 2005. Biomass power generation: sugarcane bagasse and trash. Série Caminhos para Sustentabilidade. Piracicaba: PNUD-CTC, Brazil.

IBGE, 2005. Indicadores Agropecuários 2005. Instituto Brasileiro de Geografia e Estatistica. Available at: http://www.ibge.org.br.

IBGE, 2006. Produção Agrícola Municipal, v. 33. Instituto Brasileiro de Geografia e Estatistica. Available at: http://www.ibge.org.br.

IEA, 2006. Alternative Energy Scenarios. World Energy Outlook 2006. International Energy Agency.

IPCC, 2006. IPCC guidelines for national greenhouse gas inventories, Prepared by the National Greenhouse Gas Inventories Programme. In: H.S. Eggleston, L. Buendia, K. Miwa, T. Ngara and K. Tanabe (eds.) Japan: IGES.

Macedo, I.C. (ed.) 2005. Sugarcane's Energy – Twelve studies on Brazilian sugarcane agribusiness and its sustainability. São Paulo: Berlendis & Vertecchia: UNICA, Brazil.

Macedo, I.C., 2008. GHG mitigation and cost analyses for expanded production and use of fuel ethanol in Brazil. Final Report, prepared for Center for Clean Air Policy – CCAP. Washington, USA.

Macedo, I.C., J.E.A. Seabra and J.E.A.R. Silva, 2008. Green house gases emissions in the production and use of ethanol from sugarcane in Brazil: The 2005/2006 averages and a prediction for 2020. Biomass and Bioenergy, 32: 582-595.

MAPA, 2007. Projeções do Agronegócio – Mundo e Brasil, 2006/07 a 2017/18. Ministério da Agricultura, Pecuária e Abastecimento (MAPA), Assessoria de Gestão Estratégica, Brazil.

Nassar, 2008. Sustainability considerations for ethanol. Food, Fuel and Forests: A Seminar in Climate Change, Agriculture and Trade. Bogor, Indonesia.

Nassar, A.M., B.F.T. Rudorff, L.B. Antoniazzi, D. Alves de Aguiar, M.R.P. Bacchi and M. Adami, 2008. Prospects of the sugarcane expansion in Brazil: impacts on direct and indirect land use changes. In: P. Zuurbier and J. van de Vooren (eds.) Sugarcane ethanol, Wageningen Academic Publishers, Wageningen, the Netherlands, pp. 63-93.

Seabra, J.E.A., 2008. Avaliação técnico-econômica de opções para o aproveitamento integral da biomassa de cana no Brasil. Campinas, Faculdade de Engenharia Mecânica, Universidade Estadual de Campinas, PhD Thesis, 273 p.

Scolari, D., 2006. Produção Agricola Mundial: o potencial do Brasil. Embrapa (Empresa Brasileira de Pesquisa Agopecuária), Brasilia, Brazil.

UNICA, 2008. Frequently asked questions about the Brazilian sugarcane industry. UNICA, S Paulo, Brazil.

Chapter 5
Environmental sustainability of sugarcane ethanol in Brazil

Weber Antônio Neves do Amaral, João Paulo Marinho, Rudy Tarasantchi, Augusto Beber and Eduardo Giuliani

1. Introduction

Brazil's economy is performing well during the last few year reaching international investment grade levels, while at the same time providing quantifiable reductions of greenhouse gases, specially through its renewable energy matrix and the large scale use of ethanol in transportation. It is well-known that the quality of life in the world increases with economic growth, which increases demand for energy (Figure 1). If one considers the externalities created by burning fossil fuels, then economic growth becomes a major threat to the global well being; reinforcing the need to explore alternatives to improve the efficiency of energy use and diversification of energy sources, and especially from renewable ones.

Brazil's commitment to sustainability in the agribusiness for example can be assessed by concrete examples such as the development and implemental of stringent legal environmental frameworks, agricultural zoning, massive investments in research and development and rural social policies, being the ethanol business a good example from which best practices could be disseminated.

The benefits of the production and use of ethanol in Brazil can also serve as a platform and model for further acceptance and deployment of renewable sources of biomass as feedstock for sustainable production of biofuels in the World. However there are several drivers that currently affect the supply and demand for biofuels and their sustainable production: land use changes, environmental concerns, competition with other sources of energy, food security, agricultural subsidies, innovation and technological development, public policies, oil prices, energy security policies, etc.

The Proalcool program (the Brazilian program for the production of ethanol) started in 1975, 33 years ago, is a good example of a pro-active public policy supporting the development of biofuels with a focus on sugarcane ethanol. It made Brazil the second largest producer of ethanol (expected production of 23 billion liters in 2008), with the lowest production costs in the World (US$ 0.22/l – Table 1).

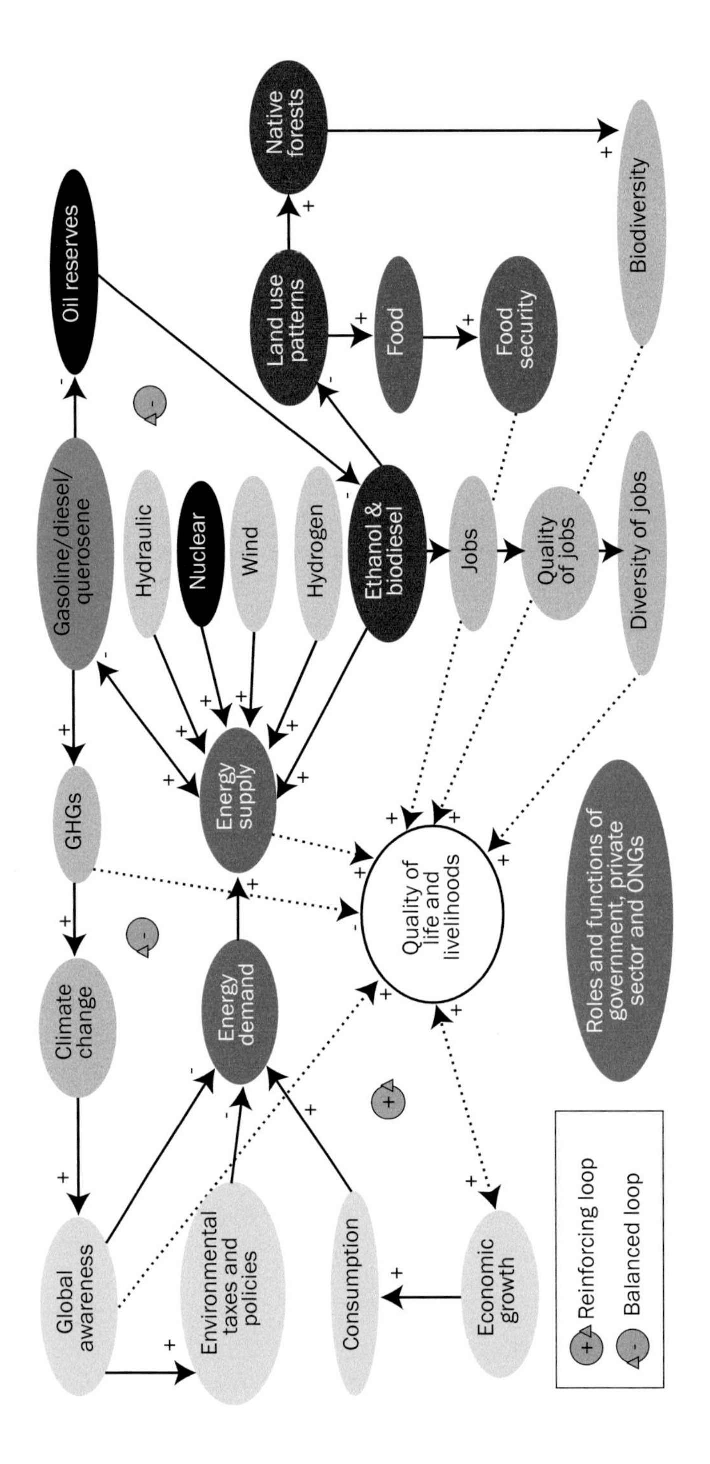

Figure 1. Social-economical-environmental dynamic structure. Source: Venture Partners do Brasil.

Table 1. Production costs of different biofuels (US$/liter)[1].

Costs item	United States							Brazil	EU
	Maize wet milling	Maize dry milling	Sugarcane	Sugar beets	Molasses [3]	Raw sugar [3]	Refined sugar [3]	Sugarcane [4]	Sugarbeets [4]
Feedstock[2]	0.11	0.14	0.40	0.43	0.25	0.84	0.97	0.08	0.26
Processing	0.17	0.14	0.25	0.21	0.10	0.10	0.10	0.14	0.52
Total	0.28	0.28	0.65	0.63	0.34	0.94	1.07	0.22	0.78

Source: USDA (2007).

[1] Excludes capital costs.

[2] Feedstock costs for US maize wet and dry milling are net feedstock costs; feedstock costs for US sugarcane and sugar beets are gross feedstock costs.

[3] Excludes transportation costs.

[4] Average of published estimates.

The long track record of Brazilian sugarcane ethanol proved its economic sustainability over time, while improving its social and environmental indicators, involving technology transfer from Europe, US and other regions and developing several innovations at national level. This program no longer exists, however it has contributed significantly to improve the productivity of sugarcane and ethanol extraction rates (Figure 2).

Due to the increasing internal demands and the possibility of future exports, it is expected that the Brazilian production might increase to 47 billion liters of ethanol by 2015, with an estimated annual growth rate of 10-13% (Table 2).

Several steps will be necessary to achieve these production targets, including sustainable planning of the sugarcane expansion into new areas, improving the logistics, the development of global markets and continuously developing new technological innovations, while at the same time improving the environmental performance of existing brown fields (areas with already established sugarcane fields and industry either/or sugar mills/distilleries) and especially from new green fields (new areas for expansion of sugarcane fields and new industrial plants), which are being implemented using cutting edge technologies in the agriculture and in the industry. With more than 360 mills in operation, there is a gap between the best practices available and the average performance of Brazilian mills, however due to recent developments in the ethanol business, with the consolidation of economic groups, capacity building programs, companies going public, new investments

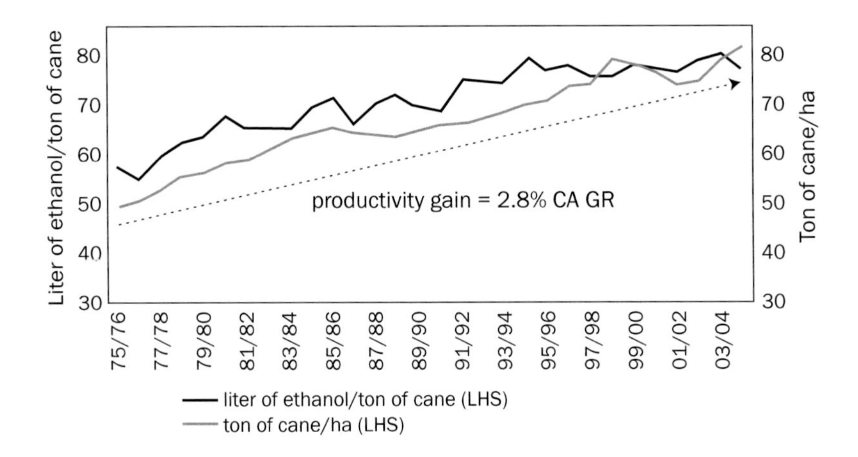

Figure 2. Evolution of productivity of Brazilian ethanol. Source: Itaú Corretora (2007).

Table 2. Future projections of ethanol production in Brazil.

	2007/08	2015/16	2020/21
Sugarcane Production (M-ton)	493	829	1,038
Area (M-ha)	7.8	11.4	13.9
Sugar (M-ton)	30.8	41.3	45
Internal	12.2	11.4	12.1
Export	18.6	29.9	32.9
Ethanol (B-liters)	22.5	46.9	65.3
Internal	18.9	34.6	49.6
Export	3.6	12.3	15.7
Bioelectricity (GW average)	1.8	11.5	14.4

Source: Unica (2008).

in research and development the speed of this dissemination is increasing significantly from previous decades.

This chapter addresses the following:
- the Brazilian environmental legal frameworks;
- key environmental indicators: carbon, water, soil, agrochemicals, biodiversity, air and by-products;
- different biofuels certifications regimes and compliance;
- the future steps and the role of innovation.

2. The Brazilian environmental legal framework regulating ethanol production

The Brazilian environmental legal framework is complex and one of the most stringent and advanced in the World. As an agribusiness activity, the ethanol/sugar industry has several environmental restrictions that require appropriate legislation or general policies for its operation. Some of them are pioneers in the area which define principles in order to maintain the welfare of living beings and to provide resources for future generations: the first version of the Brazilian forest code dated from 1931, already addressed the need to combine forest cover with quality of life and livelihoods.

Brazil has wide range of federal and state laws regarding environmental protection (Table 3), aiming at combining the social economic development with environmental preservation, which the ethanol business need to comply with for its proper operation.

They also involve frameworks such as the Environmental Impact Assessment and Environmental Licensing, among others (Figure 3), especially for the implementation of new project: i.e. new green field projects in Brazil are being stringently assessed (Nassar *et al.*, this book) using these frameworks.

Volunteer adherence to Environmental Protocols represents also a major breakthrough for the sugar business. For example The 'Protocolo Agroambiental do Setor Sucroalcooleiro' (Agriculture and Environmental Protocol for the ethanol/sugar industry) signed by UNICA and the Government of the State of São Paulo in June 2007 deals with issues such as: conservation of soil and water resources, protection of forests, recovery of riparian corridors and watersheds, reduction of greenhouse emissions and improve the use of agrochemicals and fertilizers. But its main focus is anticipating the legal deadlines for ending sugarcane burning by 2014 from previous deadline of 2021. In February 2008, the State Secretariat of Environment reported that 141 industries of sugar and alcohol had already signed the Protocol, receiving the 'Certificado de Conformidade Agroambiental' (Agricultural and Environmental Certificate of Compliance). These adherences correspond for more than 90% of the total sugarcane production in São Paulo. A similar initiative is happening in the State of Minas Gerais with the 'Protocolo de Intenções de Eliminação da Queima da Cana no Setor Sucroalcooleiro de Minas Gerais' from August 2008.

Table 3. Summary of main environmental laws.

Law	Objective	P.S.
No. 4,771, September 15th, 1965	Forest Code	Permanent preservation areas
No. 997, May 31st, 1976	Environment Polution Control	Environmental Permission
Portaria do Ministério do Interior No. 323, November 29th, 1981	It prohibits release of vinhoto in the water	
No. 6,938, August 31st, 1981	Environment National Policy	Mechanisms and instruments (environmental zoning, Environmental Impact Assessment)
CONAMA deliberation No.001/7986	General Guidelines for the Evaluation of Environmental Impact	For 'industrial complex anc units and agro-industrial'
No. 6,171, July 04th, 1988	The use, conservation and preservation of agricultural soil	
No. 11,241, September 19th, 2002	Gradual elimination of burning the straw of sugarcane	Elimination of the use of fire as a unstraw method and facilitator of cutting the sugarcane
No. 12183/05	Use of water charge	
No. 50,889, June 16th, 2006	Legal Reserve of landed property in the State of São Paulo	Obligation of reserving an area equivalent to 20% of each rural property
SMA deliberation 42, October 14th, 2006	Environmental prior license to distilleries of alcohol, sugar plants and units of production of spirits	It defines criteria and procedures
Deliberation No. 382, December 26th, 2006	It sets the maximum emission of air pollutants to sources.	Annex III: Emission limits for air pollutants from processes of heat generation from the external combustion of sugarcane's mulch
Agricultural and Environmental Protocol of sugar/ethanol industry	Prominence to anticipate the legal period to the end of the harvest of sugarcane with the previous use of fire in the areas cultivated by plants	Government of the State of São Paulo and UNICA
Elimination intentions of burning sugarcane in the ethanol/sugar sector of Minas Gerais protocol	Removal of burnt by 2014	SIAMIG/SINDAÇÚCAR-MG and Government of the State of Minas Gerais

Source: Brazilian and State laws.

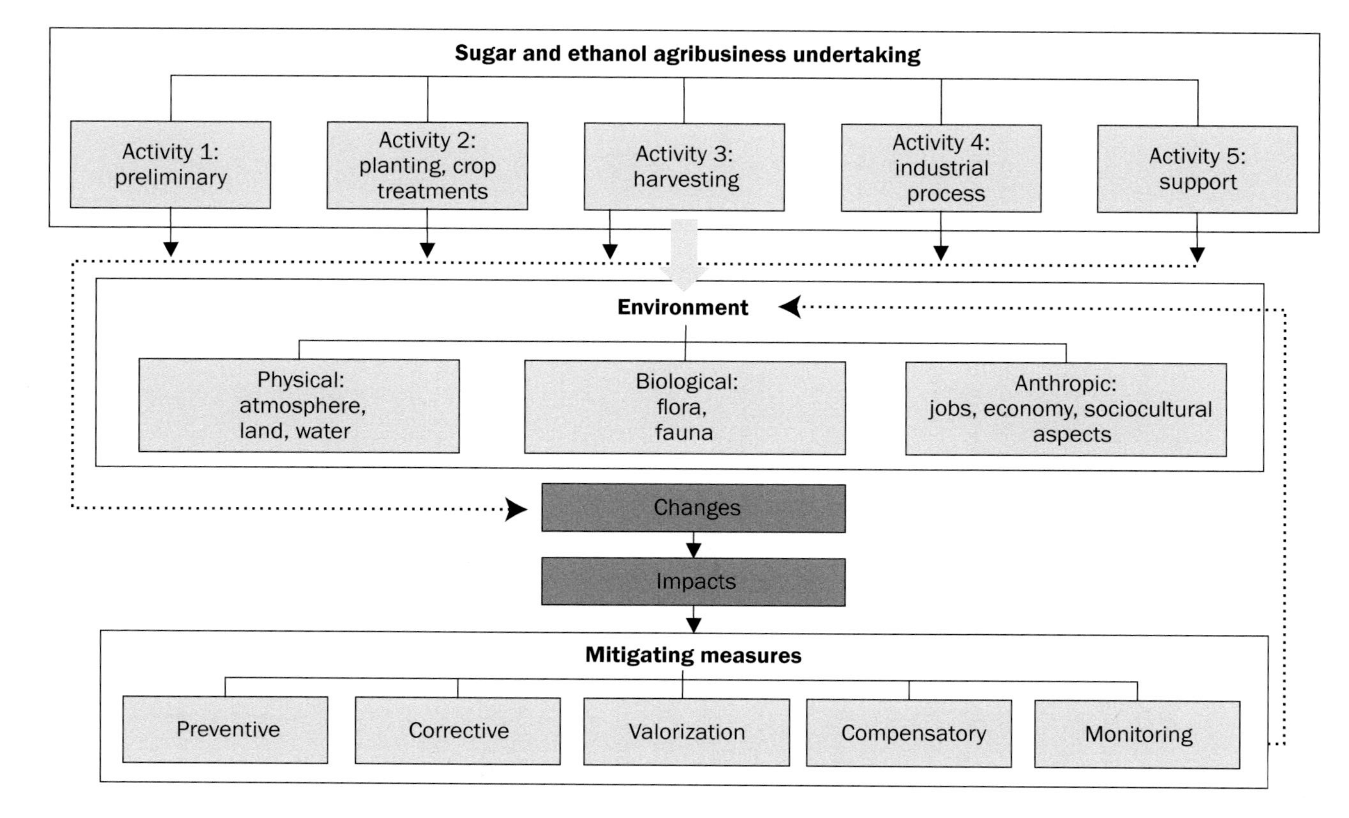

Figure 3. Environmental analysis structure for EIA/RIMA: sugarcane agribusiness. Source: Elia Neto (2008).

3. Environmental indicators

The environmental sustainability is evaluated through indicators such as carbon, water, soil, agrochemicals, biodiversity and by-products.

3.1. Greenhouse gases (GHG) balance

One of the goals of using biofuels is to contribute with net reduction of GHG emissions and thus not affecting carbon stock negatively in different sub-systems of production, below and above ground biomass (roots, branches and leaves) and in the soil (carbon fixed in clay, silt, sand and organic matter). Figure 4 shows that ethanol from sugarcane reduces 86% of the GHG emissions when compared to gasoline. It has also a leading performance when compared to other biofuels from other feedstocks. In addition the energy efficiency difference is even greater: 9.3 against 1.4 to 2.0 of other biomasses (Figure 5).

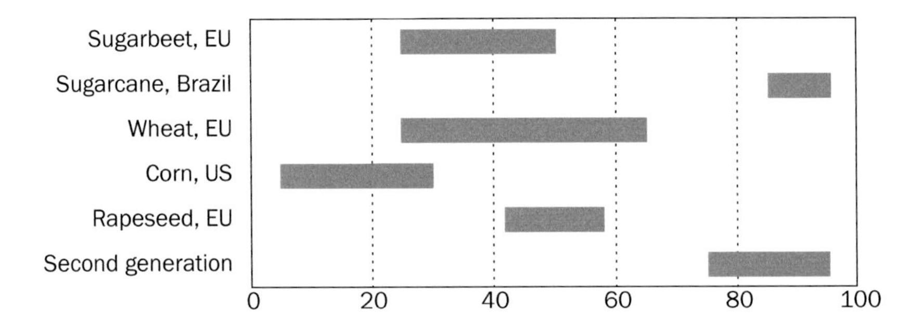

Figure 4. GHG emissions avoided with ethanol or biodiesel replacing gasoline. Source: International Energy Agency (IEA/OECD, 2006).

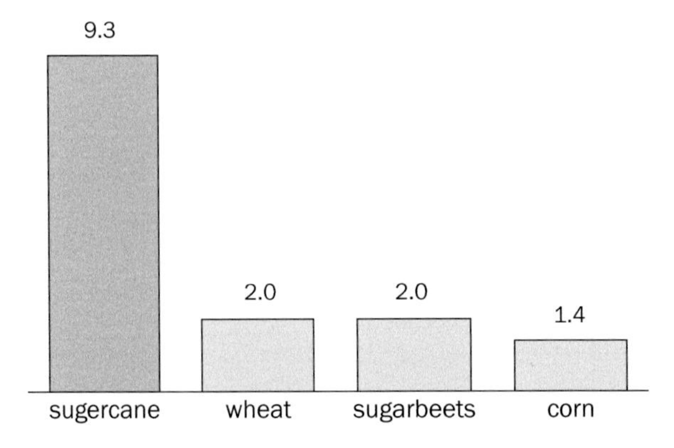

Figure 5. Energy output per unit of fossil fuel consumption in the production process. Source: World Watch Institute (2006) and Macedo *et al.* (2008).

3.1.1. Carbon stocks

One of the main effects caused by land use changes is the variation in the amount of carbon stocks under different subsystem, namely in the soil and in the above ground biomass in the area. When analyzing the environmental effects caused by different land use regimes, the balance of carbon should be taken into account. It is necessary to know how much carbon would be fixed or released into the air under different land use regimes compared with the previous baseline of use.

One limiting factor to perform an in depth analysis of these balances is the lack of long term monitoring plots assessing precisely these dynamics through time. However the stock and flows of carbon for major crops like soybean, maize, cotton and sugarcane have been extensively studied, but in general using different methodologies. There are also other factors that affect the results: crop productivity and management, soil physical and chemical properties, climate and land use history for example.

In large countries such as Brazil, there are many different soils and climatic conditions. The different characteristics of each region will influence the potential for carbon storage. A clay soil, for example, has the ability to store more organic matter and consequently, more carbon than a sandy soil, because of their physical properties. In hot and humid climates, the rate of deposition and decomposition of organic matter is higher than in dry and cold climates, facilitating the deposition of carbon in the soil.

The spatial distribution of crops in Brazil is edaphic-climatic (soil characteristics and climate interactions) dependent for their profitability. These interactions influence carbon content in the soil and in the biomass, which are also affected by soil management practices, such as minimum tillage, which can significantly for example increase soil carbon content. The land use history is also relevant when assessing and explaining current levels of carbon, because when land use changes do occur; soil carbon stocks take several years to achieve a new carbon balance. If carbon is measured in a newly cultivated system, the carbon present in the soil is actually reflecting the carbon content from the formerly existing vegetation/ history and not a consequence of current land use. Table 4 presents the carbon stocks in soil for some selected Brazilian crops and in the native vegetation.

For carbon stored in the biomass, crop productivity is of great importance as indicator carbon stored in the above ground biomass per unit of area. The larger the quantity of biomass above ground, the greater the stocks of carbon in biomass (Table 5), which is a measure much easier to obtain and with a larger dataset from multiple management and production systems in Brazil.

According to the National Supply Company (CONAB - Ministry of Agriculture and Livestock, 2008) sugarcane area expanded 653,722 ha in the 2007/2008 period, occupying

Table 4. Carbon stock in soil for selected crops in native vegetation.

Biomass	**Carbon stocks in soil (Mg/ha)**
Campo Limpo - grassland savannah (a)	72
Sub-tropical forest (b)	72
Tropical forest (c)	71
Natural pasture (d)	56
Soybean (e)	53
Cerradão - woody savannah (a)	53
Managed pasture (f)	52
Cerrado - typical savannah (a)	46
Sugarcane without burn (g)	44
Degraded pasture (h)	41
Maize (h)	40
Cotton (i)	38
Sugarcane burned (g)	35

Sources: (a) Lardy *et al.* (2001); (b) Cerri *et al.* (1986); (c) Trumbore *et al.* (1993); (d) Jantalia *et al.* (2005); (e) Campos (2006); (f) Rangel and Silva *et al.* (2007); (g) Estimated from Galdos (2007); (h) d'Andréa *et al.* (2004); (i) Neves *et al.* (2005).

Table 5. Carbon stocks in the above biomass of selected crops and native vegetation.

Biomass	**Carbon stocks in biomass (Mg/ha)**
Tropical rain forest (a)	200.0
Cerradão - woody savannah (b)	33.5
Cerrado - typical savannah (b)	25.5
Sugarcane without burn (c)	17.5
Sugarcane burned (c)	17.0
Campo Limpo - grasland savannah (b)	8.4
Managed pasture (d)	6.5
Maize (e)	3.9
Cotton (f)	2.2
Soybean (g)	1.8
Degraded pasture (d)	1.3

Sources: (a) INPE; (b) Ottmar *et al.* (2001); (c) VPB Estimative; (d) Estimated from Szakács *et al.* (2003); (e) Estimated from Titon *et al.* (2003); (f) Adapted from Fornasieri and Domingos *et al.* (1978); (g) Adapted from Campos (2006).

areas previously covered with pasture (67%), soybean (16.9%), maize (4.9%) and 2.4% of these new areas expanded into native vegetation of cerrado (savannah-like vegetation). From these numbers, it is possible to estimate the overall carbon balance resulting from land use changes due to sugarcane expansion for this period (Table 6). Figure 6 shows the positive carbon balance resulting from 91.2% in the area of expansion of sugarcane, corresponding to the areas of pasture, maize, soybeans and native vegetation as replaced by not burned sugarcane as 100% of these new green field are using mechanized harvesting practices. It was considered for this assessment that the totality of pastures replaced was of planted pastures and the native vegetation replaced as areas of Grassland Savannah (Campo Limpo). However it is important to mention that there are other statistics of sugarcane expansion (See Nassar *et al.* in this volume for details), which could affect this carbon balance.

Table 6. Carbon balance under different land uses replaced by sugarcane.

Biomass	**Total carbon stocks (Mg/ha)**	**Carbon balance due to sugarcane replacement (Mg/ha)**
Cotton (d)	40.1	21.8
Degraded pasture (b)	42.0	19.8
Maize (h)	44.1	17.7
Sugarcane burned (g)	52.1	9.7
Soybean (e)	54.9	6.9
Managed pasture (f)	58.5	3.3
Cerrado - typical savannah (a)	71.5	-9.7
Campo Limpo - grassland savannah (a)	80.4	-18.6
Cerrado - woody savannah (a)	86.5	-24.7
Tropical forest (c)	271.0	-209.2
Total carbon stocks in sugarcane net burned = 61.8 Mg/ha		

Sources: (a) Lardy *et al.* (2001)/Ottmar *et al.* (2001); (b) d`Andréa *et al.* (2004)/Estimated from Szakács *et al.* (2003); (c) Trumbore *et al.* (1993)/INPE; (d) Neves *et al.* (2005)/Adapted from Fornasieri and Domingos *et al.* (1978); (e) Campos (2006)/Adapted from Campos (2006); (f) Rangel and Silva *et al.* (2007)/Estimated from Szakács *et al.* (2003); (g) Estimated from Galdos (2007)/ VPB Estimative; (h) d`Andréa *et al.* (2004)/Estimated from Titon *et al.* (2003).

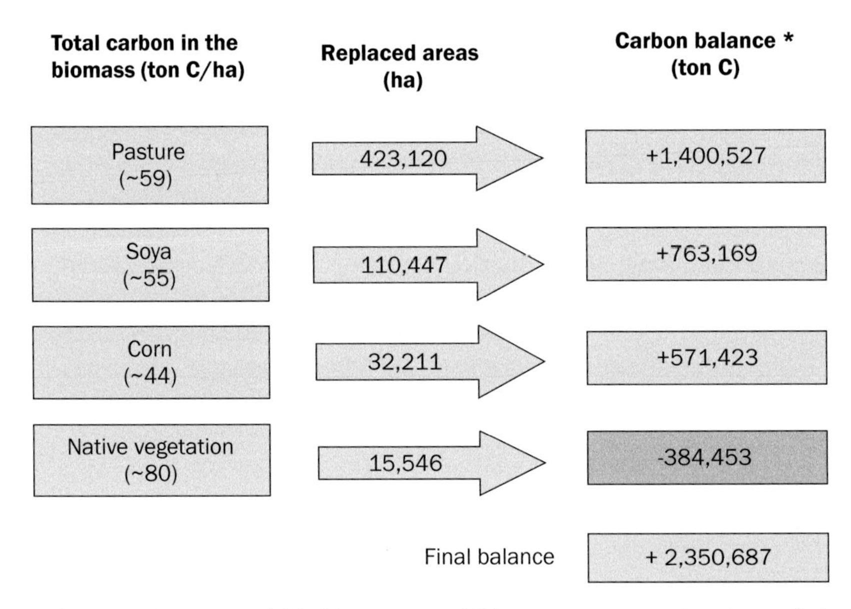

Figure 6. Carbon balance of sugarcane expansion in São Paulo State, 2007. Source: VPB analysis.

3.2. Water

Despite having the greatest water availability in the world, with 14 percent of the surface waters and the equivalent of annual flow in underground aquifers, the use of crop irrigation in Brazil is minimum (~3.3 Mha, compared to 227 Mha in the world). Practically all of the sugarcane produced in São Paulo State is grown without irrigation (Donzelli, 2005).

The levels of water withdraw and release for industrial use have substantially decreased over the past few years, from around 5 m^3/ton sugarcane collected in 1990 and 1997 to 1.83 m^3/ton sugarcane in 2004 (sampling in São Paulo). If we take 1.83 m^3 of water/ton of sugarcane, and exclude the mills having the highest specific consumption, the mean rate for the mills that account for 92% of the total milling is 1.23 m^3 of water/ton of sugarcane. In addition the recycling rate has been increasing since 1990 (Figure 7). Mills with better water management practice replace only 500 liters in the industrial system, with a recycling rate of 96,67%.

Recent developments might lead to convert sugarcane mills from water consumers to water exporters industry. Dedini the largest Brazilian manufacturer of sugar mills and equipment suppliers has developed a new technology that allows the process of transforming sugarcane in ethanol to be much more efficient, and in the end of this process, industrial mills will be able to sell about 300 liters of water per ton of sugarcane (Figure 8). This would be

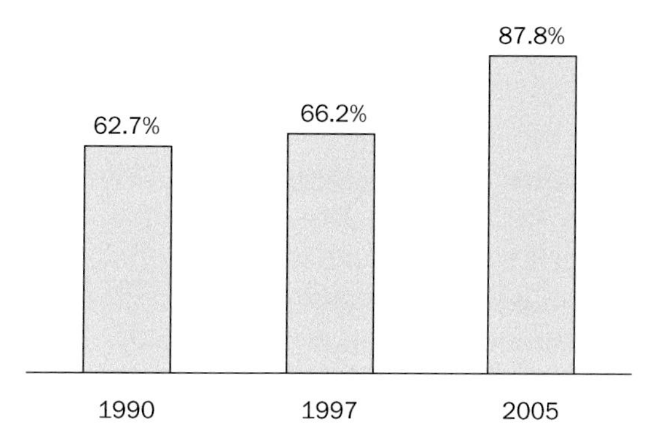

Figure 7. Evolution of water recycling. Source: Elia Neto (2008).

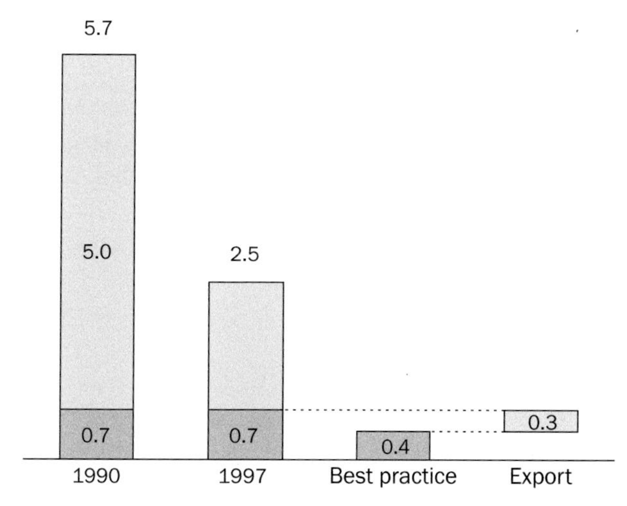

Figure 8. Evolution of water consumption in industrial ethanol production from sugarcane (m^3/ton of sugarcane). Source: Dedini (2008).

possible because water represents approximately 70% of sugarcane's composition. This new technology will be available next year (2009). Current estimates from maize ethanol mills on water consumption are of 4 liters of water per liter of ethanol produced (Commission on Water Implications of Biofuels Productions in United States, 2008).

3.3. *Soil and fertilizers*[8]

The sustainability of the culture improves with the protection against soil erosion, compacting and moisture losses and correct fertilization. In Brazil, there are soils that have been producing sugarcane for more than 200 years, with ever-increasing yields and soil carbon content. Soil erosion in sugarcane fields is lower than in soybean and maize (Macedo *et al.*, 2005) and other crops (Table 7). It is expected also that the growing harvesting of cane without burning will further improve this condition, with the use of the remaining trash in the soil. Recent sugarcane expansion in Brazil has happened mostly in low fertility soils (pasture lands), and thus improving their organic matter and nutrient levels from previous land use patterns. Sugarcane uses lower inputs of fertilizers: ten, six and four times lower than maize respectively for nitrogen, phosphorous and potassium (Table 8). An important characteristic of the Brazilian sugarcane ethanol is the full recycling of industrial waste to the field.

Vinasse, a by-product of the distillation process, rich in nutrients (mainly potassium) and organic matters is a good example, which is being used extensively as a source of ferti-irrigation (nutrients associated with water). For each liter of alcohol, 10 to 15 liters of vinasse

[8] This text was adapted from Donzeli (2005) and Souza (2005).

Table 7. Losses of soil and water for selected crops.

Annual crop	Losses	
	Soil (t/ha-year)	**Water (% rain)**
Castor	41.5	12.0
Beans	38.1	11.2
Manioc	33.9	11.4
Peanut	26.7	9.2
Rice	25.1	11.2
Cotton	24.8	9.7
Soybean	20.1	6.9
English potato	18.4	6.6
Sugarcane	12.4	4.2
Maize	12.0	5.2
Maize + beans	10.1	4.6
Sweet potato	6.6	4.2

Source: Bertoni *et al.* (1998).

Table 8. Agrochemical inputs consumption (per ha) and per ethanol production (m^3).

	Sugarcane		Maize	
	Cons./ha	Cons./m^3	Cons./ha	Cons./m^3
Ethanol production (m^3)	8.1	-	4.2	-
Quantity of N (kg)	25.0	3.1	140.0	33.7
Quantity of P (kg)	37.0	4.6	100.0	24.1
Quantity of K (kg)	60.0	7.4	110.0	26.5
Liming materials (kg)	600.0	74.5	500.0	120.5
Herbicide (liters)	2.6	0.3	13.0	3.1
Drying hormone (liters)	0.4	0.0	-	-
Insecticides (liters)	0.1	0.0	2.2	0.5
Formicide (kg)	-	-	0.5	0.1
Nematicide (liters)[a]	1.2	0.1	-	-
Total	726.2	90.2	865.7	208.5

Sources: Agrianual (2008); Fancelli and Dourado Neto (2006).
[a] Product used to control microscopic multicellular worms called nematodes.

are produced. Generally the vinasse has a high organic matter and potassium content, and relatively poor nitrogen, calcium, phosphorus and magnesium contents (Ferreira and Monteiro 1987). Advantages of using vinasse include increased pH and cation exchange capacity, improved soil structure, increased water retention, and development of the soil's micro flora and micro fauna. Many studies have been conducted involving specific aspects pertaining to leaching and underground water contamination possibilities at variable vinasse doses over periods of up to 15 years. The results obtained from tests so far indicate that there are no damaging impacts on the soil at doses lower than 300 m^3/ha, while higher doses may damage the sugarcane or, in specific cases (sandy or shallow soil), contaminate underground water (Souza, 2005).

Investments in infrastructure have enabled the use water from the industrial process and the ashes from boilers. Filter cake (a by product of the yeast fermentation process) recycling processes were also developed, thereby increasing the supply of nutrients to the field.

3.4. Management of diseases, insects and weeds[9]

Strategies for disease control involve the development of disease resistant varieties within large genetic improvement programs. This approach kept the major disease outbreak managed, i.e. the SCMV (sugarcane mosaic virus, 1920), the sugarcane smut, *Ustilago scitaminea*, and rust *Puccinia melanocephala* (1980's), and the SCYLV (sugarcane yellow leaf virus, 1990's) by replacing susceptible varieties.

The soil pest monitoring method in reform areas enabled a 70% reduction of chemical control (data provided by CTC), thereby reducing costs and risks to operators and the environment.

Sugarcane, as semi-permanent culture of annual cycle and vegetative propagation, forms a crop planted with a certain variety that is reformed only after 4 to 5 years of commercial use. These characteristics determine that the only economically feasible disease control option is to use varieties genetically resistant to the main crop diseases.

Insecticide consumption in sugarcane crops is lower than in citrus, maize, coffee and soybean crops; the use of insecticides is also low, and of fungicides is virtually null (Agrianual, 2008). Among the main sugarcane pests, the sugarcane beetle, *Migdolus fryanus* (the most important pest) and the cigarrinha, *Mahanarva fimbriolata*, are biologically controlled. The sugarcane beetle is the subject of the country's largest biological control program. Ants, beetles and termites are chemically controlled. It has been possible to substantially reduce the use of pesticides through selective application.

The control or management of weeds encompasses specific methods or combinations of mechanical, cultural, chemical and biological methods, making up an extremely dynamic process that is often reviewed. In Brazil, sugarcane uses more herbicides than coffee and maize crops, less herbicides than citrus and the same amount as soybean (Agrianual, 2008).

On these issues mentioned above related to use of agrochemicals, soil management and water uses, UNICA's (Brazilian Sugarcane Growers Association) associated mills are developing a set of goals, aiming at improving agricultural sustainability in the next few years (Table 9).

---

[9] This text was adapted from Arrigoni and Almeida (2005) and Ricci Junior (2005).

Table 9. Sugarcane agricultural sustainability.

Sugarcane		
Less agrochemicals	**Low soil loss**	**Minimal water use**
Low use of pesticides. No use of fungicides Biological control to mitigate pests. Advanced genetic enhancement programs that help idntify the most resistant varieties of sugarcane. Use of vinasse and filter cake as organic fertilizers.	Brazilian sugarcane fields have relatively low levels of soil loss, thanks to the semi-perennial nature of the sugarcane that is only replanted every 6 years. The trend will be for current losses, to decrease significantly in coming years through the use of sugarcane straw, some of which is left on the fields as organic matters after mechanical harvesting	Brazilian sugarcane fields require practically no irrigation because rainfall is abundant and reliable, particularly in the main South Central production region. Ferti-irrigation: applying vinasse (a water-based residue from sugar and ethanol production). Water use during industrial processing has decreased significantly over the years: from 5 m^3/t to 1 m^3/t.

Source: Unica (2008).

3.5. Conservation of biodiversity

Brazil is a biodiversity hotspot and contains more than 40% of all tropical rain forest of the World. Brazilian biodiversity conservation priorities were set mainly between 1995 and 2000, with the contribution of hundreds of experts; protected areas were established for the six major biomes in the National Conservation Unit System.

Steps for the implementation of the Convention on Biological Diversity includes the preparation of the biodiversity inventory and monitoring of important biodiversity resources, the creation of reserves, the creation of seed, germoplasm and zoological banks, and the conduct of Environmental Impact Assessments covering activities that could affect the biodiversity.

The percentage of forest cover represents a good indicator of conservation of biodiversity in agricultural landscapes. In São Paulo State for example the remaining forest covered is 11%, of which 8% being part of the original Atlantic Forest. Table 10 demonstrates that while the sugarcane area increased from 7 to 19% of the State territory, native forests also increased from 5 to 11%, showing that it is possible to recover biodiversity in intense agricultural systems.

Table 10. Sugarcane and vegetation area in São Paulo State.

Year	Sugarcane					Vegetation			% SP State	
	New lands (Kha)	Land in use (Kha)	Total area (Kha)	Production (Kton)	Productivity (ton/ha)	Woody-Cerradao (Kha)	Shrubby-Cerrado/ savana (Kha)	Native forests (Kha)	Sugarcane area	Native forests area
1983	345	1,421	1,765	107,987	76.0	196	489	1,139	7%	5%
1984	317	1,526	1,842	116,666	76.5	167	427	1,453	7%	6%
1985	326	1,626	1,952	121,335	74.6	221	438	1,545	8%	6%
1986	350	1,704	2,054	122,986	72.2	205	378	1,795	8%	7%
1987	311	1,753	2,064	132,322	75.5	211	348	1,870	8%	8%
1988	325	1,771	2,097	134,108	75.7	192	316	1,624	8%	7%
1989	322	1,757	2,078	130,795	74.5	198	325	1,487	8%	6%
1990	276	1,836	2,112	139,400	75.9	175	290	1,097	9%	4%
1991	301	1,864	2,165	144,581	77.6	198	301	1,601	9%	6%
1992	372	1,940	2,311	150,878	77.8	204	284	2,109	9%	8%
1993	371	1,989	2,360	156,623	78.7	238	259	2,120	10%	9%
1994	421	2,180	2,601	168,362	77.2	201	238	2,453	10%	10%
1995	449	2,260	2,709	175,073	77.5	189	220	2,434	11%	10%
1996	428	2,388	2,816	187,040	78.3	217	232	2,462	11%	10%
1997	422	2,451	2,872	194,801	79.5	215	244	2,478	12%	10%
1998	342	2,544	2,887	199,764	78.5	217	241	2,482	12%	10%
1999	281	2,475	2,756	193,374	78.1	218	244	2,468	11%	10%
2000	338	2,491	2,829	189,391	76.0	221	257	2,629	11%	11%
2001	440	2,569	3,009	201,683	78.5	223	262	2,622	12%	11%
2002	457	2,661	3,118	212,707	79.9	224	263	2,725	13%	11%
2003	495	2,818	3,313	227,981	80.9	225	264	2,720	13%	11%
2004	463	2,951	3,414	241,659	81.9	211	262	2,732	14%	11%
2005	553	3,121	3,673	254,810	81,7	217	254	2,648	15%	11%
2006	822	3,437	4,258	284,917	82,9	228	271	2,695	17%	11%
2007	935	3,897	4,832	327,684	84,1	233	277	2,716	19%	11%

Source: IEA/CATI-SAAESP (Annual statistics from 1983-2007).

3.6. Air quality

Burning sugarcane for harvesting is one of the most criticized issue of sugarcane production system, causing local air pollution and affecting air quality, despite of the benefits of using 100% ethanol running engines instead of gasoline (Figure 9), which decreases air pollution from 14 to 49%.

In order to eliminate gradually sugarcane burning, several attempts are being made. The São Paulo Green Protocol is being considered the most important one, setting an example for other regions and states in Brazil. Signed between the São Paulo state government (State Environment Secretariat) and the Sugarcane Growers Association (UNICA) in June 04, 2007, the Green Protocol aimed at:

- The anticipation of the legal deadline for the elimination of the practice of sugarcane straw burning to 2014.
- The protection of river side woods and recovering of those near water streams (permanent protected areas - APPs).
- The implementation of technical plans for conservation of soil and water resources.
- The adoption of measures to reduce air pollution.
- The use of machines instead of fire to harvest new sugarcane fields.

Voluntarily 141 of the total of 170 sugar mills from the state of São Paulo signed this Protocol, and recently 13 thousand sugarcane independent suppliers, members of the Organization of Sugarcane Farmers of the Center-South Region (Orplana), signed also this protocol. Therefore the entire production chain of sugar and ethanol of São Paulo participates

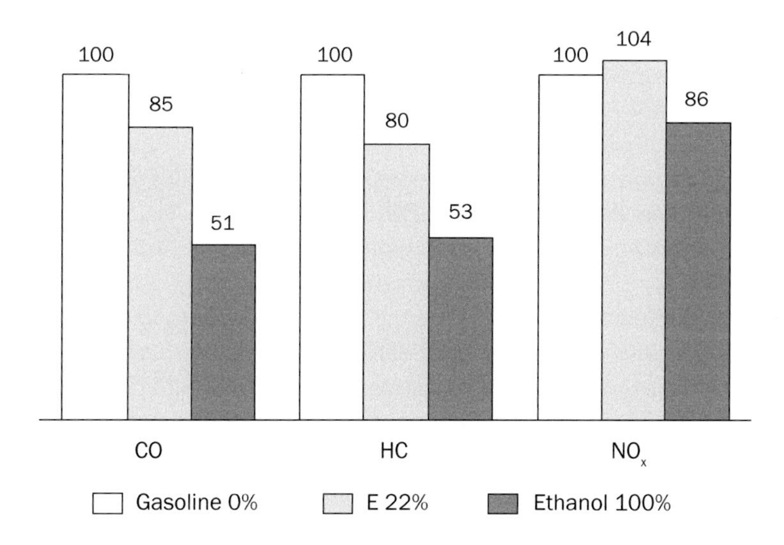

Figure 9. Air pollution by different blends of ethanol. Source: ANFAVEA (2006).

in the implementation of the Protocol. Maintaining the 2007 levels of mechanization, when 550 new harvest machines have begun to operate, it will be possible to complete the mechanization even prior to the deadline (2014) set by the Protocol.

4. Initiatives towards ethanol certification and compliance

The discussion on sustainable production of biofuels has fulfilled the scientific literature lately (see for example Hill *et al.*, 2006; Van Dam *et al.*, 2006; Goldemberg *et al.*, 2006; Smeets *et al.*, 2008; Macedo *et al.*, 2008). At the same time several initiatives are being developed in Europe and in the United States related to certification, traceability and definition of criteria and indicators for sustainable production of biofuels, mainly due to different supporting policies. For example in May 2003, the European Commission launched its Biofuels Directive 2003/30/EC, establishing legal basis for blending biofuels and fossil fuels. The EU member countries are urged to replace 2% of fossil fuels with biofuels by 2005 and 5.75% by 2010. From 2003 to 2005 the group of 25 countries members enhanced biofuel's market share of 0.6% to 1.4%. However, they have not yet achieved the first target yet. The EU Directive 2003/96/EC had also established tax incentives to encourage renewable energy use.

The government of Germany (GE), Netherlands (NL) and United Kingdom (UK) are supporting different assessment studies, while another one initiative is taking place from Switzerland, the Roundtable on Sustainable Biofuels (RTB), a multiple stakeholder initiative, hosted by the Ecole Polytechnique Federale de Lausanne. The main environmental issues addressed by these different initiatives are related to greenhouse gas reduction compared with fossil fuels; competition with other land uses, especially food competition; impacts on the biodiversity and on the environment (Table 11). Considering carbon and greenhouse gases balance current agricultural and industrial practices sugarcane ethanol from Brazil does comply with the targets of greenhouse reduction higher than 79% from existing brown fields, and from new green fields, when not replacing large areas of native vegetation. On food competition, there is no direct evidence that sugarcane is replacing the basic Brazilian staple foods (Nassar *et al.*, this book). On biodiversity conservation, data from São Paulo State show that sugarcane expansion did not reduce forest cover, but on the contrary (IEA/ CATI – SAAESP). On the use of water, fertilizers and agrochemicals, sugarcane ethanol does perform well above any other current biofuel in the market (in this chapter).

In the USA, the Environmental Protection Agency (EPA) under the Energy Independence and Security Act of 2007 is responsible for revising and implementing regulations on the use of biofuels blended with gasoline. The Renewable Fuel Standard program will increase the volume of renewable fuel required to be blended into gasoline from 9 billion gallons in 2008 to 36 billion gallons by 2022. At the same time, EPA is conducting several studies on the direct and indirect impacts of the expansion of biofuels production and their carbon footprint and potential reduction of greenhouse gases.

Table 11. Main issues related to sustainable production of biofuels being considered under different certification regimes.

Criterion	NL	UK	GE	RTB	EU
1. Greenhouse gas balance	✓	✓	✓	✓	✓
a1) Net emission reduction compared with a fossil fuel reference is at least 50%. Variation in policy instruments could benefit the best performances.	✓				
a2) Life cycle GHG balance reduction of 67% compared with fossil fuels			✓		
a3) Processing of energy crops GHG reduction of 67% compared with fossil fuels			✓		
a4) GHG emissions savings from the use of biofuels at least 35% compared with fossil fuels		✓			✓
a5) GHG emissions will be reduced when compared to fossil fuels				✓	
b) Soil carbon and carbon sinks		✓		✓	✓
c) Emissions of N_2O from biofuels		✓			
2. Competition with other applications/ land use	✓	✓	✓	✓	✓
a) Availability of biomass for food, local energy supply, building materials or medicines should not decline	✓	✓	✓	✓	✓
b) Use of less productive land for biofuels		✓			
c) Increasing maximum use of crops for both food and fuel		✓			
d) Avoiding negative impacts from bioenergy-driven changes in land use			✓	✓	
3. Biodiversity	✓	✓	✓	✓	✓
a) No deterioration of protected area's or high quality eco-systems.	✓	✓	✓	✓	✓
b) Insight in the active protection of the local ecosystem.	✓				
c) Alteration of local habitats		✓			
d) Effect on local species		✓		✓	
e) Pest and disease resistance		✓			
f) Intellectual property and usage rights	✓		✓		
g) Social circumstances of the local residents	✓		✓		
h) Integrity	✓				
i) Standard on income distribution and poverty-reduction			✓		
j) Avoiding human health impacts			✓		
4. Environment	✓	✓	✓	✓	
a) No negative effects on the local environment					
b) Waste management	✓	✓			
c) Use of agro-chemicals, including artificial manure	✓	✓	✓		

Table 11. Continued.

Criterion	NL	UK	GE	RTB	EU
4. Environment (continued)	✓	✓	✓	✓	
d) Preventing erosion and deterioration of the soil to occur and maintaining the fertility of the soil	✓	✓	✓	✓	
e) Active improvement of quality and quantity of surface and groundwater	✓		✓	✓	
f) Water use efficiency of crop and production chain		✓	✓		
g) Emissions to the air	✓			✓	
h) Use of genetically modified organisms			✓	✓	

Source: adapted from Van Dam *et al.* (2006).
NL = the Netherlands; UK = United Kingdom; GE = Germany; RTB = Round table on sustainable biofuels; EU = European Union.

While the above concerns are well-justified, some criticism of biofuels and their impacts are motivated by protectionism and interest in agricultural subsidies and agribusiness production chains in several developing countries, especially from EU countries. Certification schemes suggested may become non-tariff barriers, rather than environmentally and socially sound schemes.

Scientific and technological assessments comparing different kinds of biofuels are needed to reduce the play of such interests and to establish the strengths of best potential of biofuels along with their dangers and limitations.

The OECD's latest report on biofuels illustrates how fears can be perpetuated without proper scientific basis. Suggestively titled: ('Biofuels: is the cure worse than the disease?'), the report stated: 'Even without taking into account carbon emissions through land-use change, among current technologies only sugarcane-to-ethanol in Brazil, ethanol produced as a by-product of cellulose production (as in Sweden and Switzerland), and manufacture of biodiesel from animal fats and used cooking oil, can substantially reduce [greenhouse gases] compared with gasoline and mineral diesel. The other conventional biofuel technologies typically deliver [greenhouse gas] reductions of less than 40% compared with their fossil-fuel alternatives'.

This report also recognized that while still trade barriers would persist to the international market, it will be difficult for the world to take advantage of the environmental qualities of the use of some biofuels, mainly the ethanol form sugarcane and so forth as international markets are not yet fully created for biofuels.

5. Future steps towards sustainable production of ethanol and the role of innovation

A huge challenge facing policy makers, businesses, scientists and societies as a whole is how to responsibly establish sustainable production systems and biofuel supplies in sufficient volume that meet current and future demands globally.

The examples and best practices found in Brazilian sugarcane ethanol provides a good framework and baseline of sustainability compared with other current biofuels available in large scale in the World, having the smallest impact on food inflation, high levels of productivity (on average 7,000 liters of ethanol/ha and 6.1 MWhr of energy/ha), with lower inputs of fertilizers and agrochemicals, while reducing significantly the emissions of greenhouse gases. The ending of sugarcane burning in 2014 is a good example of improving existing practices. The proper planning of sugarcane expansion into new areas will for another important step towards sustainable production of ethanol

In addition new technologies and innovation are taking place in Brazil and elsewhere in the world, aiming at optimizing the use of feedstocks: using lignocellulosic materials (the second generation of biofuels); reducing waste; adding value to ethanol co-products and moving towards ethanol chemistry and biorefinaries full deployment.

Different initiatives in Brazil from the State of São Paulo Research Foundation (FAPESP), Ministry of Science and Education (MC&T - FINEP) and investments from the private sector are contributing to the deployment of new opportunities provided by the sugarcane biomass, at the same time improving the environmental performances at the agriculture and at the industry.

References

Agrianual, 2008. Anuário da Agricultura Brasileira. FNP Consultoria & Agroinformativo, São Paulo, 502 pp.

Arrigoni, E. De Beni and L.C. De Almeida, 2005. Use of agrochemicals. In: Macedo, I.C. (ed.) Sugarcane's Energy - Twelve studies on Brazilian sugarcane agribusiness and its sustainability. São Paulo, pp. 147-161.

Associação Nacional dos Fabricantes de Veículos Automotores (ANFAVEA), 2006. Brazil. Available at: www.anfavea.com.br.

Bertoni, J., F.I. Pastana, F. Lombardi Neto and R. Benatti Junior, 1998. Conclusões gerais das pesquisas sobre conservação de solo no Instituto Agronômico. In: Lombardi Neto, F. and Bellinazi Jr, R. (eds.) Simpósio sobre terraceamento agrícola, Campinas, SP, Fundação Cargill, Campinas. Campinas, Instituto Agronômico, Jan 1982, Circular 20, 57 pp.

Brazilian Government. Lei No. 6.938, de 31 de agosto de 1981. Dispõe sobre a Política Nacional do Meio Ambiente, seus fins e mecanismos de formulação e aplicação, e dá outras providências. Available at: www.planalto.gov.br/ccivil_03/Leis/L6938.htm.

Brazilian Government. Lei No. 4.771, de 15 de setembro de 1965. Institui o novo Código Florestal. Available at: www.planalto.gov.br/ccivil_03/LEIS/L4771.htm.

Brazilian Government. Portaria do Ministério do Interior n. 323, de 29 de novembro de 1978. Proíbe lançamento de vinhoto em coleções de água. Brazil. Available at: www.udop.com.br/download/legislacao_vinhaca.pdf.

Brazilian Government. Resolução CONAMA n. 001 de 23 de janeiro de 1986. Estabelecer as definições, as responsabilidades, os critérios básicos e as diretrizes gerais para uso e implementação da Avaliação de Impacto Ambiental como um dos instrumentos da Política Nacional do Meio Ambiente. Available at: www.mma.gov.br/port/conama/res/res86/res0186.html. Brazil.

Campos, B.C., 2006. Dinâmica do carbono em latossolo vermelho sob sistemas de preparo de solo e de culturas. Santa Maria, RS. Tese (doutorado), Universidade Federal de Santa Maria.

Cerri, C.C., 1986. Dinâmica da matéria orgânica do solo no agrossistema cana-de-açúcar. Tese (livre docência). Escola Superior de Agricultura 'Luiz de Queiroz', Piracicaba, SP, Brazil.

Committee on Water Implications of Biofuels Production in the United States, 2008. Water Implications of Biofuels Production in the United States. The National Academies Press. 88 pp.

Companhia Nacional De Abastecimento (CONAB), 2008. Perfil do Setor do Açúcar e do Álcool no Brasil. Brasília: Conab, 2008. Available at: http://www.conab.gov.br/conabweb/download/safra/perfil.pdf.

D'Andrea, A.F., M.L.N. Silva, N. Curi and L.R.G. Guilherme, 2004. Estoque de carbono e nitrogênio e formas de nitrogênio mineral em um solo submetido a diferentes sistemas de manejo. Pesquisa agropecuária. brasileira, Brasília, vol. 39 (2): 179-186.

Dedini, 2008. Dedini lança usina de açúcar e etanol produtora de água. Press release. Dedini S.A. Indústria de Base, Piracicaba.

Donzelli, J.L., 2005. Preservation of agricultural soil. In: Macedo, I.C. (ed.) Sugarcane's Energy – Twelve studies on Brazilian sugarcane agribusiness and its sustainability. São Paulo, pp. 139-146.

Elia Neto, A., 2008. Água na indústria da Cana-de-açúcar. In: Workshop Projeto PPP: 'Aspecto Ambientais da Cadeia do Etanol de Cana-de-açúcar', Painel I, São Paulo. 13 pp.

Fancelli, A.L. and D. Dourado Neto, 2004. Produção de milho. 2. ed. Guaíba, Agropecuária, Brazil. 360 pp.

Ferreira, E.S. and A.O. Monteiro, 1987. Efeitos da aplicação da vinhaça nas propriedades químicas, físicas e biológicas do solo. Boletim Técnico Copersucar 36: 3-7.

Fornasieri, F. and V.I. Domingos, 1978. Nutrição e adubação mineral do algodoeiro. In: Ministério da Ciência e Tecnologia – MCT (ed.). Emissões de gases de efeito estufa na queima de resíduos agrícolas.Brasília/DF – MCT, 2002. Available at: www.mct.gov.br/index.php/content/view/17341.html.

Galdos, M.V., 2007. Dinâmica do carbono do solo no agrossistema cana-de-açúcar. Tese (doutorado) – Escola Superior de Agricultura 'Luiz de Queiroz', Piracicaba.

Goldemberg, J., 2007. Ethanol for a Sustainable Energy Future. Science 315: 808-810.

Hill, J., E. Nelson, D. Tilman, S. Polasky and T. Douglas, 2006. Environmental, economic, and energetic costs and benefits of biodiesel and ethanol biofuels. PNAS 103: 11206-11210.

IEA/CATI – SAAESP, 2008. Área cultivada e produção: Cerrado, Cerradão, Mata Natural. In: Banco de Dados do Instituto de Economia Agrícola. São Paulo. Available at: www.iea.sp.gov.br.

Instituto de Economia Agrícola (IEA), 2008. Available at: www.iea.sp.gov.br.

Instituto Nacional de pesquisas espaciais (INPE), 2008. Available at: www.inpe.br.

International Energy Agency, 2004-2006. United States. Available at: www.iea.org.

Itaú Corretora, 2007. Cosan: Céu Carregado, Tese de investimentos. São Paulo.

Jantalia, C.P., 2005. Estudo de sistemas de uso do solo e rotações de culturas em sistemas agrícolas brasileiros: dinâmica de nitrogênio e carbono no sistema solo – planta – atmosfera. Tese (doutorado) – Universidade Federal Rural do Rio de Janeiro, Rio de Janeiro.

Lardy, L.C., M. Brossard, M.L.L. Assad and J.Y. Laurent, 2002. Carbon and phosphorus stocks of clayey ferrassols in Cerrado native and agroecossystems, Brazil. Agriculture Ecosystems and Environment 92: 147-158, 2002.

Macedo, I.C., J.E.A. Seabra and J.E.A.R. Silva, 2008. Green house gases emissions in the production and use of ethanol from sugarcane in Brazil: The 2005/2006 averages and a prediction for 2020. Biomass and Bioenergy 32: 582-595.

Macedo, I.C., 2005. Sugarcane's Energy – Twelve studies on Brazilian sugarcane agribusiness and its sustainability. São Paulo. Available at: http://english.unica.com.br/multimedia/publicacao/

Ministry of Agriculture and Livestock - CONAB, 2008. www.conab.gov.br. Brazil.

Neves, C.S., V.J. Feller and Larré-Larrouy, M.-C., 2005. Matéria orgânica nas frações granulométricas de um latossolo vermelho distroférrico sob diferentes sistemas de uso e manejo. Ciências Agrárias, Londrina 26: 17-26.

OECD, 2007. Biofuels: is the cure worse than the disease? Paris: September, 2007. Available at: www.oecd.org/dataoecd/40/25/39266869.pdf.

Ottmar, R.D., R.E. Vihnanek, H.S. Miranda, M.N. Sato and S.M.A. Andrade, 2003. Séries de estéreo-fotografias para quantificar a biomassa da vegetação do cerrado no Brasil Central. Brasília: USDA, USAID, UnB, 2001. In: Ciclagem de Carbono em Ecossistemas Terrestres – O caso do Cerrado Brasileiro. Planaltina/,DF – Embrapa Cerrados, 2003. Available at: bbeletronica.cpac.embrapa.br/2003/doc/doc_105.pdf.

Protocolo Agroambiental do Setor Sucroalcooleiro, 2008. Available at: www.unica.com.br/content/show.asp?cntCode={BEE106FF-D0D5-4264-B1B3-7E0C7D4031D6}.

Rangel, O.J.P. and C.A. Silva, 2007. Estoque de carbono e nitrogênio e frações orgânicas de latossolo submetido a diferentes sistemas de uso e manejo. Revista Brasileira de Ciência do Solo 31: 1609-1623.

Ricci Junior, A., 2005. Pesticides: herbicides. In: Macedo, Isaias de Carvalho; Several Authors, 2005. Sugarcane's Energy – Twelve studies on Brazilian sugarcane agribusiness and its sustainability. São Paulo, pp 156-161.

São Paulo State Government. Decreto no. 50.889, de 16 de junho de 2006. Dispõe sobre a manutenção, recomposição, condução da regeneração natural e compensação da área de Reserva Legal de imóveis rurais no Estado de São Paulo e dá providências correlatas. São Paulo. Available at: www.cetesb.sp.gov.br/licenciamentoo/legislacao/estadual/decretos/2006_Dec_Est_50889.pdf.

São Paulo State Government. Lei no. 997, de 31 de maio de 1976. Dispõe sobre o controle da poluição do meio ambiente. São Paulo. Available at: www.ambiente.sp.gov.br/uploads/arquivos/legislacoesambientais/1976_Lei_Est_997.pdf.

São Paulo State Government. Lei no. 6.171, de 04 de julho de 1988. Dispõe sobre o uso, conservação e preservação do solo agrícola. São Paulo. Available at: sigam.ambiente.sp.gov.br/Sigam2/legisla%C3%A7%C3%A3o%20ambiental/Lei%20Est%201988_06171.pdf.

São Paulo State Government. Lei no. 11.241, de 19 de Setembro de 2002. Dispõe sobre a eliminação gradativa da queima da palha da cana-deaçúcar e dá providências correlatas. São Paulo. Available at: sigam.ambiente.sp.gov.br/sigam2/Legisla%C3%A7%C3%A3o%20Ambiental/Lei%20Est%202002_11241.pdf.

São Paulo State Government. Lei no. 12.183, de 29 de dezembro de 2005. Dispõe sobre a cobrança pela utilização dos recursos hídricos do domínio do Estado de São Paulo, os procedimentos para fixação dos seus limites, condicionantes e valores e dá outras providências. São Paulo. Available at: www.ana.gov.br/cobrancauso/_ARQS-legal/Geral/Legislacoes%20Estaduais/SP/Lei-12183 05.pdf.

São Paulo State Government. Resolução SMA 42, de 14 de outubro de 2006. Licenciamento ambiental prévio de destilarias de álcool, usinas de açúcar e unidades de fabricação de aguardente. São Paulo. Available at: www.milare.adv.br/ementarios/legislacao2006.pdf. São Paulo State Government. Resolução no. 382, de 26 de dezembro de 2006. Estabelece os limites máximos de emissão de poluentes atmosféricos para fontes fixas. São Paulo. Available at: www.mp.rs.gov.br/areas/ambiente/arquivos/boletins/bola_leg01_07/ib382.pdf.

Smeets, E., M. Junginger, A. Faaij, A. Walter, P. Dolzan and W. Turkenburg, 2008;The sustainability of Brazilian ethanol - an assessment of the possibilities of certified production. Biomass and Bioenergy 32: 781-813.

Souza, S.A.V. de, 2005. Impacts on the water supply. In: Macedo, Isaias de Carvalho; Several Authors, 2005. Sugarcane's Energy - Twelve studies on Brazilian sugarcane agribusiness and its sustainability. São Paulo, pp. 105-118.

Szakács, G.G.J., 2003. Avaliação das potencialidades dos solos arenosos sob pastagens, Anhembi - Piracicaba/SP. Piracicaba, 2003. Dissertação (mestrado) - Centro de Energia Nuclear na Agricultura.

Titon, M., C.O. Da Ros, C. Aita, S.J. Giacomini, E.B. Do Amaral and M.G. Marques, 2003. Produtividade e acúmulo de nitrogênio no milho com diferentes épocas de aplicação de N-uréia em sucessão a aveia preta. In: XXIX Congresso Brasileiro de Ciência do Solo, 2003, Ribeirão Preto - SP.

Trumbore, S. 1993. Comparison of carbon dynamics in tropical and temperate soils using radiocarbon measurements. Global Biogeochemical Cycles, Washington, v.7, pp. 75-290. In: Silveira, A.M., Victoria, R.L., Ballester, M.V., De Camargo, P.B., Martinelli, L.A. and Piccolo, M.C., 2000. Simulação dos efeitos das mudanças do uso da terra na dinâmica de carbono no solo na bacia do rio Piracicaba. Revista Pesquisa. agropecuária. brasileira, Brasília 35: 389-399.

União da Indústria de Cana-de-açúcar (UNICA), 2008. Available at: www.unica.com.br.

Unica, 2008. Protocolo Agroambiental do Setor Sucroalcooleiro. Available at: www.unica.com.br/content/show.asp?cntCode={BEE106FF-D0D5-4264-B1B3-7E0C7D4031D6}.

Unica, 2008. Estimativa da Safra 2008/2009. Available at: www.unica.com.br/noticias/show.asp?nwsCode={8ACECE47-BD9D-4D82-9BFF-5ABCCD9C5953}.

United States Department of Agriculture (USDA). Available at: www.usda.gov.

Van Dam, J., M. Junginger, A. Faaij. I. Jurgens, G. Best and U. Fritsche, 2006. Overview of recent developments in sustainable biomass certification. IEA Bioenergy Task 40. 40 pp.

Venture Partners do Brasil (VPB), 2008. VPB analysis of sugarcane future scenarios. São Paulo, SP, Brazil, 14 pp.

World Watch Institute, 2006. Available at: www.worldwatch.org.

Chapter 6
Demand for bioethanol for transport

Andre Faaij, Alfred Szwarc and Arnaldo Walter

1. Introduction

The utilization of ethanol either as a straight fuel or blended to gasoline (in various proportions) has been fully proven in various countries and it is regarded as technically feasible with existing internal combustion engine technologies. Because ethanol offers immediate possibilities of partially substituting fossil fuels, it has become the most popular transport biofuel in use. Production of ethanol, which has been rising fast, is expected to reach 70 billion litres by the end of 2008. Approximately 80% of this volume will be used in the transport sector while the rest will go into alcoholic beverages or will be either used for industrial purposes (solvent, disinfectant, chemical feedstock, etc.).

Although a growing number of countries, including China and India, have been introducing ethanol in the transport fuels market, it is in Brazil, in the USA and in Sweden where this use has gained most relevance. In March 2008, consumption of ethanol surpassed that of gasoline in Brazil, largely due to the success of the flex-fuel vehicles (FFVs) and resulting steep increase in straight ethanol (E100) consumption. In the USA, in addition to a rising utilization of FFVs and high ethanol content blends with up to 85% ethanol content (E85), over 50% of the gasoline marketed now contains ethanol, mostly 10% (E10). Sweden has been leading ethanol use in Europe with the 5% gasoline blend consumed nationwide (E5), an upward demand of E85 and a fleet of 600 ethanol-fuelled buses.

The international interest on ethanol in the transport sector has been based on various reasons including energy security, trade balance, rural development, urban pollution and mitigation of global warming. The challenge for the near future is to achieve wide acceptance of ethanol as a sustainable energy commodity and global growth of its demand. In the transport sector this includes increased supply of ethanol produced from a variety of renewable energy sources in an efficient, sustainable and cost-effective way. In many countries, 2nd generation biofuels (including ethanol) produced from lignocellulosic biomass instead of food crops, is thought to deliver such performance, but commercial technology to convert biomass from residues, trees and grasses to liquid fuels is not yet available. On the demand side, it comprises the optimisation of existing engine technologies and development of new ones that could make the best possible use of ethanol and be introduced in the market in a large scale. Ethanol is a well suited and high quality fuel for more efficient flex fuel engines, ethanol-fuelled hybrid drive chains and dual-fuel combustion systems. Such technologies can boost vehicle efficiency and increase demand for ethanol use in various transport applications.

2. Development of the ethanol market

2.1. Growth in demand and production

Liquid biofuels play so far a limited role in global energy supply, and represent only 10% of total bioenergy, 1.38% of renewable energy and 0.18% of total world energy supply. They are of significance mainly for the transport sector, but even here they supplied only 0.8% of total transport fuel consumption in 2005, up from 0.3% in 1990. In recent years, liquid biofuels have shown rapid growth in terms of volumes and share of global demand for transport energy. Ethanol production is rising rapidly in many parts of the world in response to higher oil prices, which are making ethanol more competitive. In 2007 the world fuel ethanol production was estimated as 50 billion litres, being the production in USA (24.6 billion litres) and Brazil (19 billion litres) equivalent to 88% of the total; in EU the production was almost 2.2 billion litres, in China 1.8 billion litres and in Canada 800 million litres (RFA, 2008, based on Licht, 2007).

Production of ethanol via fermentation of sugars is a classic conversion route, yet the most popular, which is applied for sugarcane, maize and cereals on a large scale, especially in Brazil, the United States and to a lesser extent the EU and China. Ethanol production from food crops like maize and cereals has been linked to food price increase, although estimates to what extent vary widely and many factors apart from biofuels play a role in those price increases (FAO, 2008). In addition bioethanol from such feedstocks has only been competitive to gasoline and diesel when supported by subsidies. Despite of some advances in its production process, ethanol from food crops is not likely to achieve major cost reduction in the short and medium terms.

In contrast, the impact of sugarcane based ethanol production (dominated by Brazil) on food prices seems minimal, given reduced world sugar prices in recent years. It's production achieved competitive performance levels with fossil fuel prices without the need of subsidies (Wall-Bake *et al.*, 2008). Also it has been gaining an increasingly relevant position in other countries in tropical regions (such as India, Thailand, Colombia and various countries in Sub-Saharan Africa). Production costs of ethanol in Brazil have steadily declined over the past few decades and have reached a point where ethanol is competitive with production costs of gasoline (Rosillo-Calle and Cortez, 1998; Wall-Bake *et al.*, 2008). As a result, ethanol is no longer financially supported in Brazil and competes openly with gasoline (Goldemberg *et al.*, 2004).

Figure 1 shows the learning curves of sugarcane and ethanol from sugarcane in Brazil since late 1970s. The estimated progress ratio (PR) of 0.68 in case of sugarcane imply that its costs of production have reduced, on average, 32% each time its cumulative production has doubled (19% in case of ethanol costs, excluding feedstock costs). The figure also shows

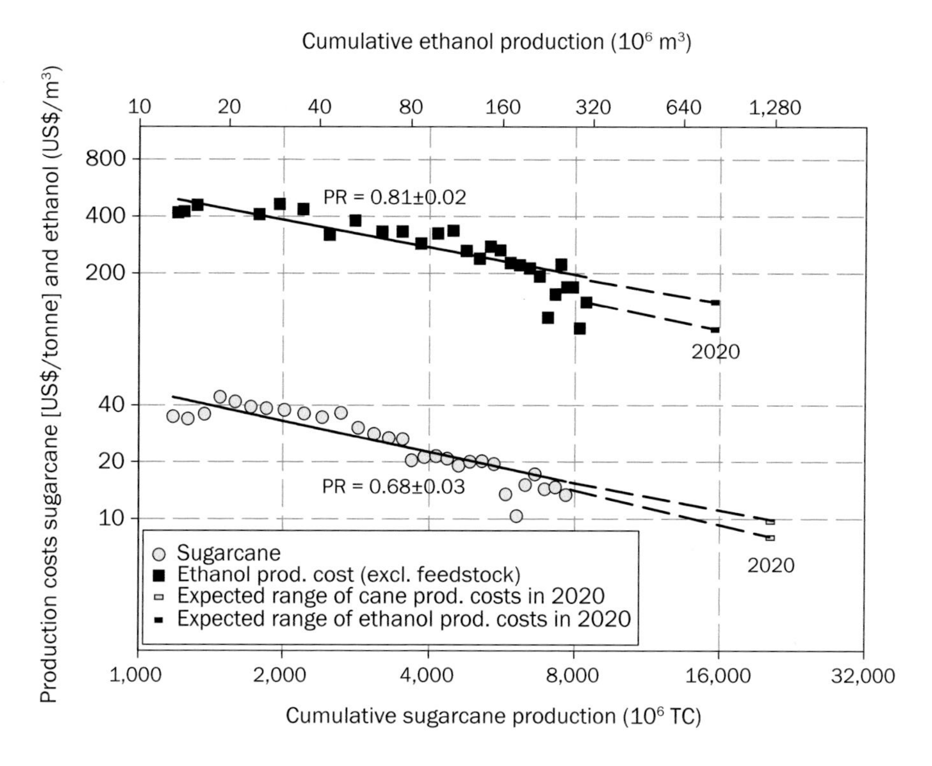

Figure 1. Learning curves and estimated future costs of sugarcane and ethanol production (excluding feedstock costs) assuming 8% annual growth of sugarcane and ethanol production (Wall-Bake *et al.*, 2008).

estimated costs of sugarcane and ethanol production by 2020, supposing a certain growth path of sugarcane and ethanol production.

Larger facilities, better use of bagasse and trash residues from sugarcane production, e.g. with advanced power generation (gasification based) or hydrolysis techniques (see below), and further improvements in cropping systems, offer further perspectives for sugarcane based ethanol production (Damen, 2001; Hamelinck *et al.*, 2005).

The growth in the use of ethanol has been facilitated by its ability to be blended with gasoline in existing vehicles and be stored and transported using current facilities, equipment and tanks. Blending anhydrous ethanol with gasoline at ratios that generally are limited to E10 has been the fastest and most effective way of introducing ethanol in the fuel marketplace.

In Brazil fuel retailers are required to market high ethanol-content blends, with a percentage that can vary from 20% to 25% by volume (E20 – E25). Vehicles are customized for these blends by car manufacturers or, in the case of imported cars (around 10% of the market), at the origin or by the importers themselves.

FFVs in the USA, Sweden and elsewhere can operate within a range that varies from straight gasoline to E85 blends, while in Brazil they are built to run in a range that varies from E20–E25 blends to E100. Up to 2006 car manufacturers in Brazil used to market dedicated E100 vehicles, which were later substituted by the FFVs.

Considering that current world's gasoline demand stands in the order of 1.2 trillion litres per year (information brochure produced by Hart Energy Consulting for CD Technologies, 2008) fuel ethanol supply will reach approximately 5% of this volume in 2008, which in energy terms represents 3% of current gasoline demand.

Ethanol has the advantage that it lowers various noxious emissions (carbon monoxide, hydrocarbons, sulphur oxides, nitrogen oxides and particulates) when compared to straight gasoline. Nevertheless the extent of emission reduction depends on a number of variables mainly engine characteristics, the way ethanol is used and emission control system features.

With regard to GHG emissions it has been demonstrated that on a life-cycle basis sugarcane ethanol produced in Brazil can reduce these emissions by 86% under current manufacturing conditions and use when compared to gasoline (Macedo *et al.*, 2008). Avoided emissions due to the use of ethanol produced from maize (USA) and wheat (EU) are estimated as 20-40% on life-cycle basis (IEA, 2004). In case of ethanol from sugarcane further reductions of GHG emissions are possible in short to mid-term, with advances in the manufacturing process (i.e. replacement of mineral diesel with biodiesel or ethanol in the tractors and trucks, end of pre-harvest sugarcane burning and capture of fermentation-generated CO_2) (Macedo *et al.*, 2008; Damen, 2001; Faaij, 2006).

2.2. International trade

The development of truly international markets for bioenergy has become an essential driver to develop available biomass resources and bioenergy potentials, which are currently underutilised in many world regions. This is true for both residues as well as for dedicated biomass production (through energy crops or multifunctional systems, such as agro-forestry). The possibilities to export biomass-derived commodities for the world's energy market can provide a stable and reliable demand for rural communities in many developing countries, thus creating an important incentive and market access that is much needed in many areas in the world. The same is true for biomass users and importers that rely on a stable and reliable supply of biomass to enable often very large investments in infrastructure and conversion capacity.

Figures 2 and 3 show the top ten ethanol importers and exporters in 2006, when the total volume traded was estimated as 6.5 billion litres, i.e. almost 13% of the whole production (Valdes, 2007). At that year more than 60 countries exported ethanol, but only ten surpassed 100 million litres traded and the most important 15 exporters covered 90% of the whole trade. US have imported more than 2.5 billion litres in 2006, EU about 690 million litres (Licht, 2007), while the imports of Japan were estimated as about 500 million litres. These three economic blocks represented about 80% of the net imports of ethanol in 2006.

Clearly, Brazil stands out as the largest exporter, covering more than 50% of the total volume traded. Except in 2006, when more than 50% was directly sold to US, ethanol exports from

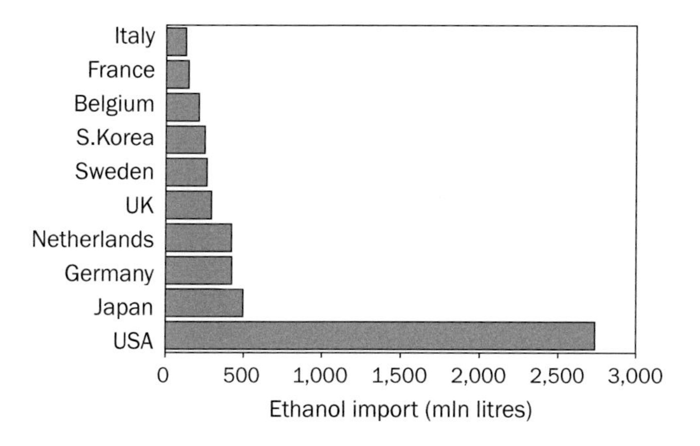

Figure 2. Top 10 ethanol importers in 2006 (Licht, 2007).

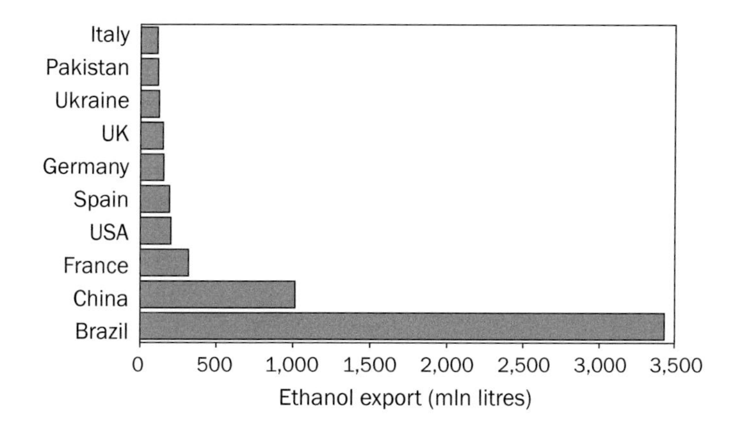

Figure 3. Top 10 ethanol exporters in 2006 (Licht, 2007).

Brazil have been roughly well distributed among 10-12 countries. On the other hand, due to the Caribbean Basin Initiative (CBI) agreement[10], most of the ethanol exported from Brazil to Central America and Caribbean countries reaches US. US importers from Caribbean and Central America countries have continuously grown since 2002.

Figure 4 shows Brazil's ethanol trade since 1970. Traditionally, Brazilian exports of ethanol have been oriented for beverage production and industrial purposes but, recently, trade for fuel purposes has enlarged. Halfway the 90-ies, a shortage of ethanol occurred, even requiring net imports. But after 2000 Brazilian exports of ethanol have risen steadily. In 2007 exports reached 3.5 billion litres and it is estimated that about 4 billion litres will be exported in 2008. It is expected that Brazil will maintain such an important position in the future. Outlooks on the future ethanol market are discussed in the next section.

[10] CBI is an agreement between US and Central American and Caribbean countries that allows that up to 7% of the US ethanol demand may be imported duty-free, even if the production itself occurs in another country (Zarilli, 2006).

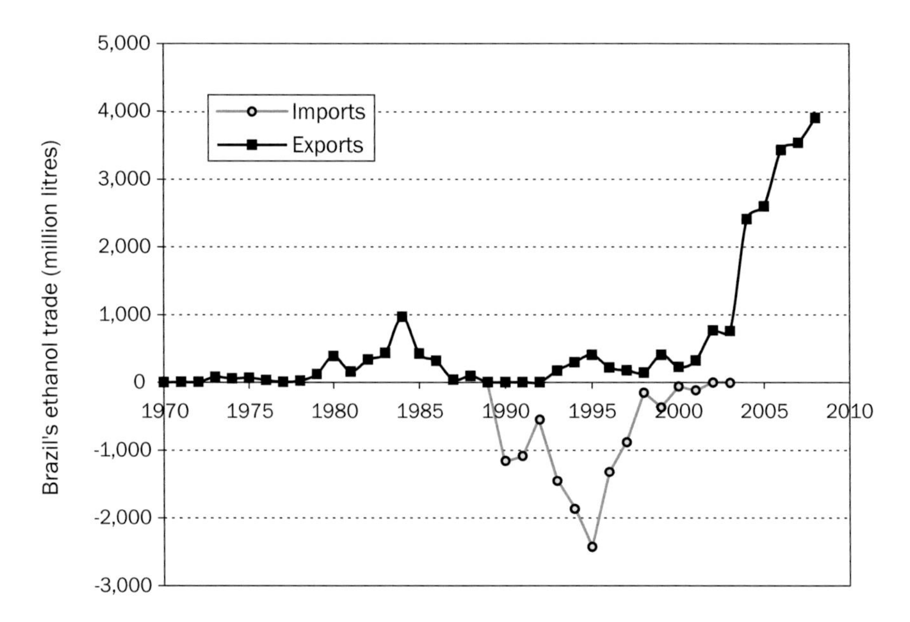

Figure 4. Trade in ethanol in Brazil 1970-2008 (estimates for 2008), including all end-uses (Brazil, 2008), (Kutas, 2008).

3. Drivers for ethanol demand

3.1. Key drivers

When evaluating key drivers for ethanol demand, energy security and climate change are considered to be the most important objectives reported by nearly all countries that engage in bioenergy development activities. As illustrated in Table 1 no country highlights less than three key objectives. This renders successful bioenergy development a challenge as it tries to reach multiple goals, which are not always compatible. For instance, energy security considerations favour domestic feedstock production (or at least diversified suppliers), whereas climate change considerations and cost-effectiveness call for sourcing of feedstocks with low emissions and costs. This implies that imports are likely to grow in importance for various industrialized countries, but also a strong pressure on developing 2nd generation biofuels that are to be produced from lignocellulosic biomass. Not surprisingly, the latter is a key policy and RD&D priority in North America and the EU.

Table 1. Main objectives of bioenergy development of G8 +5 countries (GBEP, 2008).

Country	**Objectives**						
	Climate change	**Environment**	**Energy security**	**Rural development**	**Agricultural development**	**Technological progress**	**Cost effectivenness**
Brazil	X	X	X	X	X	X	
China	X	X	X	X	X		
India			X	X		X	X
Mexico	X	X	X	X		X	
South Africa	X		X	X			
Canada	X	X	X			X	
France	X		X	X	X		
Germany	X	X		X	X	X	X
Italy	X		X		X		
Japan	X	X			X	X	
Russia	X	X	X	X	X	X	
UK	X	X	X	X			X
US	X	X	X	X	X	X	
EU	X	X	X	X	X	X	

Overall there are few differences between the policy objectives of G8 Countries and the +5 countries (Mexico, South Africa, Brazil, India, China). Rural development is more central to the +5 countries' focus on bioenergy development, and this is often aligned with a poverty alleviation agenda. Bioenergy development is also seen as an opportunity to increase access to modern energy, including electrification, in rural areas. The rural development objectives of the wealthier G8 countries focus more on rural revitalization. Similarly, in the +5 countries, agricultural objectives envisage new opportunities not just for high-end commercialised energy crop production, but also for poorer small-scale suppliers. Very important is that in many countries (both industrialized and developing) sustainability concerns, e.g. on land-use, competition with food, net GHG balances, water use and social consequences, has become an overriding issue. Development and implementation of sustainability criteria is now seen in a variety of countries (including the EU) and for various commodities (such as palm oil, sugar and soy) (Van Dam *et al.*, 2008; Junginger *et al.*, 2008).

3.2. Developments in vehicle technology

Transport predominantly relies on a single fossil resource, petroleum that supplies 95% of the total energy used by world transport. In 2004, transport was responsible for 23% of world energy-related GHG emissions with about three quarters coming from road vehicles. (see also the breakdown of energy use of different modes of transport in Table 2). Over the past decade, transport's GHG emissions have increased at a faster rate than any other energy-using sector (Kahn Ribeiro *et al.*, 2007).

Figures 5a and 5b provide projections for the growth in energy use per mode of transport and per world region. Transport activity will continue to increase in the future as economic growth fuels transport demand and the availability of transport drives development, by

Table 2. World transport energy use in 2000, by mode (Kahn Ribeiro *et al.*, 2007, based on WBCSD, 2004b).

Mode	Energy use (EJ)	Share (%)
Light-duty vehicles	34.2	44.5
2-wheelers	1.2	1.6
Heavy freight trucks	12.48	16.2
Medium freight trucks	6.77	8.8
Buses	4.76	6.2
Rail	1.19	1.5
Air	8.95	11.6
Shipping	7.32	9.5
Total	76.87	100

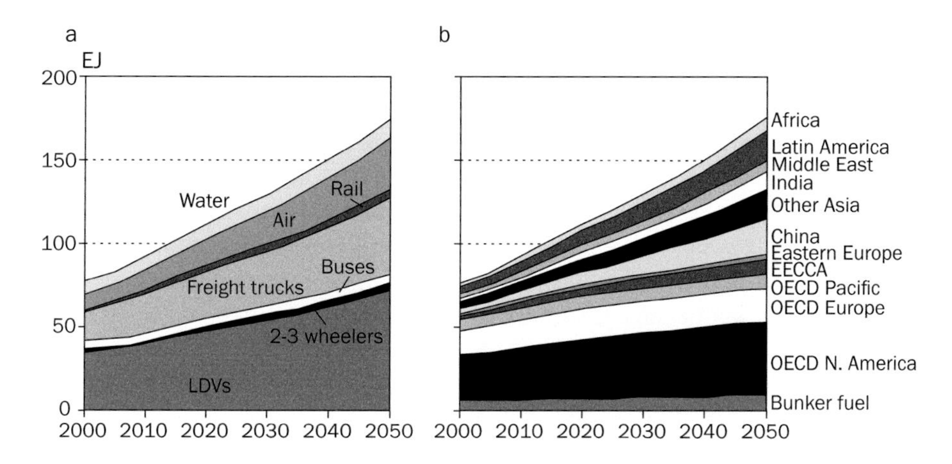

Figure 5. Projection of transport energy consumption by mode (a) and region (b) (WBCSD, 2004a).

facilitating specialization and trade. The majority of the world's population still does not have access to personal vehicles and many do not have access to any form of motorized transport. However, this situation is rapidly changing.

Freight transport has been growing even more rapidly than passenger transport and is expected to continue to do so in the future. Urban freight movements are predominantly by truck, while international freight is dominated by ocean shipping.

Transport activity is expected to grow robustly over the next several decades. Unless there is a major shift away from current patterns of energy use, world transport energy use is projected to increase at the rate of about 2% per year, with the highest rates of growth in the emerging economies. Total transport energy use and carbon emissions are projected to be about 80% higher than current levels by 2030 (Kahn Ribeiro *et al.*, 2007).

There is an ongoing debate about whether the world is nearing a peak in conventional oil production that will require a significant and rapid transition to alternative energy resources. There is no shortage of alternative energy sources that could be used in the transport sector, including oil sands, shale oil, coal-to-liquids, gas-to-liquids, natural gas, biofuels, electricity and hydrogen produced from fossil fuels or renewable energy sources. Among these alternatives, unconventional fossil carbon resources could produce competitively priced fuels most compatible with the existing transport infrastructure, but these will lead to strongly increased carbon emissions (Kahn Ribeiro *et al.*, 2007).

3.2.1. The impact of existing technologies on fuel ethanol demand

In use vehicle technologies already enable large scale use of ethanol and therefore can be considered a key driver for its worldwide use. For instance, if E10 were to become globally used today, the global FFVs fleet (estimated at 15 million vehicles as of 2008) were to use the maximum level of ethanol and 50,000 buses were equipped with dedicated ethanol engines, fuel ethanol demand would jump from current 56 billion litres to 165 billion litres, almost a 200% increase over existing demand (Szwarc, A. personal communication). The largest consumption (75%) would come from ethanol blending with gasoline.

This estimate indicates the potential demand for ethanol without any technological breakthrough and although it would not be feasible to be achieved overnight because it requires a regulatory framework and ethanol logistics, it could be gradually developed until 2020. Projections of ethanol production for Brazil, the USA and the EU indicate that supply of 165 billion litres by 2020 could be achieved with the use of a combination of first and second generation ethanol production technologies.

However, a scenario where sugarcane ethanol production in Asia, Africa, Latin America and the Caribbean could fulfil these needs is also possible. Approximately 25 million hectares of sugarcane would be needed to produce this volume worldwide using only first generation technology. With cellulosic ethanol production technologies in place using sugarcane bagasse and straw and combination of these technologies with first generation technology, the need for land use would be reduced to 20 million hectares. A third scenario considering extensive use of second generation ethanol production from various non-conventional feedstocks, including industrial residues and municipal waste, could further reduce the need of land for ethanol production further (Walter *et al.*, 2008).

3.2.2. FFVs technology and the market

In 1992, the US market saw the first commercially produced FFVs. It was a concept that would allow the gradual structuring of an ethanol market. Drivers would be allowed to run on gasoline where ethanol would not be available, therefore resolving the question on 'what comes first: the car or the fuel infrastructure?' that inhibited the ethanol market growth. Pushed by alternative energy regulations and fiscal incentives, American car manufacturers began producing FFVs that in most part ended up in government fleets. Because the number of fuel stations marketing E85 is very limited, FFVs in the US have been fuelled with straight gasoline most of the time. General Motors has been championing the FFV concept in the USA and has recently engaged in the expansion of E85 sales locations. Other companies like Ford, Chrysler and Nissan have also FFVs in their sales portfolio. By December 2008 approximately 8 million FFVs (2.8% of vehicle fleet in the US) will be on American roads but still consuming mostly gasoline (Szwarc, A., personal communication).

Sweden was the first country in Europe to start using FFVs in 1994. At first only a few imported vehicles from the US composed a trial fleet, but in 2001 FFVs sales started. In 2005 local car manufacturers like Saab and Volvo developed E85 FFVs versions. In 2007, the market share of new FFVs in Sweden was 12% and the total fleet reached 80,000 vehicles (2% of the total vehicle fleet). Over 1,000 fuel stations are selling E85 in Sweden making possible the use of E85 in FFVs. A variety of policy measures have been provided incentives for FFVs in Sweden. These include exemption of biofuels from mineral oil tax, tax benefits for companies and private car owners, free parking in 16 cities and mandatory alternative fuel infrastructure and government vehicle purchases. This initiative is part of a set of measures taken by Sweden in order to achieve its ambitious goal to be at the forefront of the world's 'green' nations and achieve a completely oil-free economy by 2020.

E100-compatible FFVs were introduced in the Brazilian market in 2003 in a different context than observed in the US or Sweden, in order to fulfil consumers' desire to use a cheaper fuel. FFVs have become a sales phenomenon and presently sales correspond to nearly 90% of new light-duty vehicle sales. All car manufacturers in Brazil have developed FFVs that are being offered as standard versions for the domestic market (over 60 models in 2008). The success of FFVs can be explained by now excellent availability of E100 and E20/E25 (at more than 35,000 fuel stations nationwide), flexibility for consumers who can choose the fuel they want depending on fuel costs and/or engine performance. Since fuel ethanol has been in general less expensive than gasoline blends (straight gasoline is not available for sale in Brazil) and gives better performance, it became the fuel of choice. Furhtermore FFV's have a 'greener' and more modern image and have higher resale value compared to conventional cars.

In 2008, the Brazilian fleet of FFVs will reach 7 million vehicles (25% of vehicle fleet) and in most cases the preferred fuel has been E100. The success of FFVs in Brazil has caught the attention of manufacturers of two wheel vehicles (motorcycles, scooters and mopeds) who are developing FFVs versions that are expected to reach the market soon.

3.2.3. The impact of new drive chain technologies

Compared to current average vehicle performance, considerable improvements are possible in drive chain technologies and their respective efficiencies and emission profiles. IEA does project that in a timeframe towards 2030, increased vehicle efficiency will play a significant role in slowing down the growth in demand for transport fuels. Such steps can be achieved with so-called hybrid vehicles which make use of combined power supply of internal combustion engine and an electric motor. Current models on the market, if optimised for ethanol use, could deliver a fuel economy of about *16 km/litre* of fuel. With further technology refinements, which could include direct injection and regenerative breaking, fuel ethanol economy of *24 km/litre* may be possible. Such operating conditions, can also deliver very low concentration of emissions.

The use of ethanol in heavy-duty diesel fuelled applications is not easy. But the well established experience with ethanol-fuelled buses in Sweden, which started in the mid-nineties, and recent research with dual-fuel use (diesel is used in combination with ethanol but each fuel is injected individually in the combustion chamber according to a preset electronically controlled engine map) indicate interesting possibilities with regard to reducing both diesel use and emissions.

Drive chain technologies that may make a considerable inroad in the coming decades, such as electric vehicles and serial hybrids, may however have a profound impact on vehicle efficiency and, to some extent, a dampening effect on the growth of transport fuel demand. Penetration of electric vehicles (cars, motorcycles and mopeds) or the use of plug-in hybrids that could be connected to the grid is still uncertain. Developments in battery technology are rapid though and electric storage capacity, charging time and power to weight ratios are continuously improved. When such improved technology is especially deployed in hybrid cars, the net effect will simply be a reduction of fuel demand. However, when deployed as plug-in hybrid, part of the fuel demand can be replaced by electricity. This could reduce the growth in demand for (liquid) transport fuels down more quickly than currently assumed in various studies.

In case Fuel Cell Vehicles (FCVs) become commercially available, this may mean a boost for the use hydrogen as fuel. Although the projected overall 'well-to-wheel' potential efficiency of e.g. natural gas to hydrogen or biomass to hydrogen for use in a FCV is very good (Hamelinck and Faaij, 2006), it is highly uncertain to what extent the required hydrogen distribution infrastructure may be available in the coming decades. Important barriers are the currently high costs of FCVs and the high investment costs of hydrogen infrastructure. Most scenarios on the demand for transport fuels towards 2030 project only a marginal role for hydrogen.

Nevertheless, the speed of penetration of such more advanced drive chains in the market and the new infrastructure they require, is uncertain and the available projections for demand of liquid transport fuels indicate that we may be looking at a doubling of demand halfway this century. Also, the overall economic and environmental performance of the use of electricity and hydrogen for transport depends heavily on the primary energy source and overall chain efficiency.

Hybrid vehicles in the transport sector and urban services seem to be at present stage a more viable alternative than FCV for the same applications. Not only is this technology more advanced in terms of commercial use but also it has many practical advantages in terms of cost and fuel infrastructure (Kruithof, 2007). Sweden has been leading the development of hybrid buses and trucks equipped with electric motor and ethanol engine. Commercial use of this type of vehicles could occur by 2010 setting a new benchmark for sustainable ethanol use.

4. Future ethanol markets

Future ethanol markets could be characterized by a diverse set of supplying and producing regions. From the current fairly concentrated supply (and demand) of ethanol, a future international market could evolve into a truly global market, supplied by many producers, resulting in stable and reliable biofuel sources. This balancing role of an open market and trade is a crucial precondition for developing ethanol production capacities worldwide.

Paramount to a solution is an orderly and defined schedule for elimination of subsidies, tariffs, import quotas, export taxes and non-tariff barriers in parallel with the gradual implementation of sustainable ethanol mandates. These measures will provide the necessary conditions to reduce risks and to attract investment to develop and expand sustainable production. Several different efforts to reach these goals are ongoing including multilateral, regional, and bilateral negotiations, as well as unilateral action. Public and private instruments such as standards, product specifications, certification and improved distribution infrastructure are important for addressing technical and sustainability issues. In addition, the development of a global scheme for sustainable production combined with technical and financial support to facilitate compliance, could ensure that sustainability and trade agendas are complementary (Best *et al.*, 2008).

4.1. Outlook on 2nd generation biofuels

Projections that take explicitly second generation options into account are more rare, but studies that do so come to rather different outlooks, especially in the timeframe exceeding 2020. Providing an assessment of studies that deal with both supply and demand of biomass and bioenergy, IPCC highlights that biomass demand could lay between 70-130 EJ in total, subdivided between 28-43 EJ biomass input for electricity and 45-85 EJ for biofuels (Barker and Bashmakov, 2007). Heat and biomass demand for industry are excluded in these reviews. It should also be noted that around that timeframe biomass use for electricity has become a less attractive mitigation option due to the increased competitiveness of other renewables (e.g. wind energy) and e.g. carbon capture and storage. (Barker and Bashmakov, 2007).

In de Vries *et al.* (2007) (based on the analyses of Hoogwijk *et al.* (2005, 2008), it is indicated that the biofuel production potential around 2050 could lay between about 70 and 300 EJ fuel production capacity depending strongly on the development scenario, i.e. equivalent to 3,100 to 9,300 billion litres of ethanol[11]. Around that time, biofuel production costs would largely fall in the range up to 15 U$/GJ, competitive with equivalent oil prices around 50-60 U$/barrel (see also Hamelinck and Faaij, 2006). A recent assessment study confirms that such shares in the global energy supply are possible, to a large extent by using perennial

[11] Based on the LHV of anhydrous ethanol (22.4 MJ/litre).

cropping systems that produced lignocellulosic biomass, partly from non-agricultural lands and the use of biomass residues and wastes. Large changes in land use and leakage effects could be avoided by keeping expanding biomass production in balance with increased productivity in agriculture and livestock management. Such a development would however require much more sophisticated policies and effective safeguards and criteria in the global market (Dornburg *et al.*, 2008).

4.2. Scenario's on ethanol demand and production

Walter *et al.* (2008) evaluated market perspectives of fuel ethanol up to 2030, considering two alternative scenarios. The first scenario reflects constrains of ethanol production in US and Europe due to the hypothesis that large-scale production from cellulosic materials would be feasible only towards the end of the period. In this case world production would reach 272,4 billion litres in 2030 (6 EJ), being only 8 billion litres of second generation ethanol, amount that would displace almost 10% of the estimated demand of gasoline.

Scenario 2 is based on the ambitious targets of ethanol production defined by US government by early 2007, i.e. consumption of about 132 billion litres by 2017. This target can only be achieved if large-scale ethanol production from cellulosic materials becomes feasible in short- to mid-term. In Scenario 2 the consumption of fuel ethanol reaches 566 billion litres in 2030 (about 13 EJ), displacing more than 20% of the demand of gasoline; 203 billion litres would be second generation ethanol.

Tables 3 summarizes results of the two scenarios for different regions/countries of the world. In case of EU, the substitution of 28.5% of gasoline volume basis (Scenario 1) would correspond to the displacement of 20% energy basis. By 2030, the estimated ethanol consumption in EU (both scenarios) and US (scenario 2) would only be possible with FFVs or even neat ethanol vehicles.

Table 3 also presents estimates of production capacity of first generation ethanol. Production capacity by 2030 was evaluated by Walter *et al.* (2008) based on the capacity available in 2005 and on projections based on trends and plans. In some cases (e.g. EU) these results were adjusted to the estimates done by the IEA (2004) as well as Moreira (2006) taking into account constraints such as land availability. It is clear that without second generation ethanol the relatively modest target to displace 10% of the gasoline demand in 2030 (Scenario 1), at reasonable cost, can only be accomplished fostering fuel ethanol production in developing countries. Second generation of ethanol would be vital if 20% of the gasoline demand is to be replaced by biofuels in 2030 (Scenario 2), although a significant contribution would have to come from conventional feedstocks mainly from developing countries.

However, the combination of lignocellulosic resources (biomass residues on shorter term and cultivated biomass on medium term) and second generation conversion technology

Table 3. Ethanol consumption by 2030 in two different scenarios and production capacity based on conventional technologies (billion litres).

Region/ country	Scenario 1	Gasoline displaced (%) [1]	Scenario 2	Gasoline displaced (%) [1]	Production capacity
US	55.3	7.4	263.7	35.0	63.0
EU	36.0	28.5	49.6	39.3	27.3
Japan	9.3	10.0	14.3	15.0	- [2]
China	21.6	10.0	33.5	15.0	18.2
Brazil	50.0	48.0[3]	50.0	48.0[3]	62.0 [4]
ROW [5]	100.2	10.0	154.9	15.0	n.c.[6]

[1] Gasoline displaced in volume basis regarding the estimated gasoline consumption in 2030.
[2] It was assumed that first generation ethanol would not be produced in Japan.
[3] Estimates of gasoline displaced considering that the substitution ratio by 2030 would be 1 litre of gasoline = 1.25 litre of ethanol. In case of Brazil there is only one scenario.
[4] In this case production capacity is not the maximum, but the capacity that should be reached considering a certain path of growth.
[5] Rest of the World.
[6] n.c. = not calculated.

offers a very strong perspective. Furthermore, sugarcane based ethanol has a key role to play at present and that role can be considerably expanded by improving the current operations further and by implementation cane based ethanol production to regions where considerably opportunities exist, especially to parts of Sub-Saharan Africa. For example, the efficient use bagasse and sugar can trash with advanced co-generation technology can increase electricity output of sugar mills considerably in various countries and thus deliver a significant contribution to (renewable) electricity production. Also, it seems realistic to assume that sugarcane based ethanol can meet the new and stringent sustainability criteria that are expected in the global market on short term (see e.g. Smeets *et al.*, 2008).

5. Discussion and final remarks

5.1. Key issues for the future markets

Biofuels in 2008 is at a crossroad: the public perception and debate have to a considerable amount pushed biofuels in a corner as being expensive, not effective as GHG mitigation option, to have insignificant potential compared to global energy use, a threat for food production and environmentally dangerous. But that basic rationale for the production and use of biofuels still stands and is stronger than ever. Climate change is accepted as a certainty, the supply of oil in relation to growing demand has developed into a strategic and economic risk, with oil prices hoovering around 130 US$/barrel at the moment of writing. Furthermore, the recent food crisis has made clear how important it is that investment and capacity building reach the rural regions to improve food production capacity and make this simultaneously more sustainable. Biofuels produced today in various OECD countries have a mediocre economic and environmental performance and many objections raised are understandable, be it overrated.

However, distinguishing those biofuels from sugarcane based ethanol production and the possibilities offered by further improvement of that production system, as well as second generation biofuels (including ethanol production from lignocellulosic resources produced via hydrolysis) is very important. It is clear though, that future growth of the biofuel market will take place with much more emphasis on meeting multiple goals, especially avoiding conflicts on land-use, water, biodiversity and at the same time achieving good GHG performance and socio-economic benefits (see e.g. Hunt *et al.*, 2007).

5.2. Future outlook

Projections for the production and use of biofuels differ between various institutions. Clearly, demand for transport fuels will continue to rise over the coming decades, also with the introduction of new drive chain technology. In fact, there could be an important synergy between new drive chains (such as serial hybrid technology) and high quality biofuels with narrow specifications (such as ethanol), because such fuels allow for optimised performance and further decreased emissions of dust and soot, sulphur dioxide and nitrous oxides.

Projections that highlight a possibly marginal role for biofuels in the future usually presume that biomass resource availability is a key constraint and that biofuel production will remain based on current technologies and crops and stay expensive (e.g. IEA, 2006, OECD/FAO, 2007). Clearly, the information compiled in this chapter shows that a combination of further improved and new conversion technologies and conversion concepts (such as hydrolysis for producing sugars of ligno cellulosic materials) and use of ligno cellulosic biomass offers a different perspective: the biomass resource basis consisting of biomass residues from forestry and agriculture, organic wastes, use of marginal and degraded lands and the possible improvement in agricultural and livestock efficiency that can release lands for additional biomass production could become large enough to cover up to one third of the global energy demand, without conflicting with food production or additional use of agricultural land. Also, the economic perspectives for such second generation concepts are very strong, offering competitiveness with oil prices equivalent to some 55 US$/barrel around 2020. Further improved ethanol production (i.e. with improved cane varieties, more efficient factories and efficiently use of bagasse and trash for power generation or more ethanol using hydrolysis processes) from sugarcane holds a similar strong position for the future.

5.3. Policy requirements and ways forward

It is very likely ethanol has a major role to play in the future worlds' energy markets. There are uncertainties though, such as dwindling public support for biofuels and possible failure to commercialise second generation technologies on foreseeable term. In case biofuels can be developed and managed to be the large and sustainable energy carriers they can in principle become (which largely depends on the above mentioned governance issues). It is also clear that sugarcane based ethanol production is one of the key systems now with a very good future outlook. In addition, ethanol is a fuel that can easily absorbed by the market. Key preconditions for achieving the sketched desirable future outlook are:

- To build on the success of current sugarcane based ethanol production and develop and implement further optimised production chains.
- Remove market barriers to allow for open trade for biofuels across the globe, while at the same time securing sustainable production by adoption of broad criteria.
- To enhance strong Research Development, Demonstration and Deployment efforts with respect to advanced, second generation conversion technologies. This concerns new, commercial stand-alone processes, but also improvements of existing infrastructure and even combinations with fossil fuels (such as co-gasification of biomass with coal for production of synfuel, combined with CO_2 capture and storage).
- To develop and broaden the biomass resources base by expanding (commercial) experience with production of woody and grassy crops. Also the enhanced use of agricultural and forestry residues can play an important role, in particular on the shorter term.
- To further develop, demonstrate and implement the deployment of broad sustainability criteria for biomass production, in general, and biofuels, in particular. This can be done by means of certification. Global collaboration and linking efforts around the globe is important now to avoid a 'proliferation of standards' and the creation of different, possible conflicting schemes.

References

Barker, T. and I. Bashmakov, 2007. Mitigation from a cross-sectoral perspective, In: B. Metz, O.R. Davidson, P.R. Bosch, R. Dave and L.A. Meyer (eds.), Climate Change 2007: Mitigation contribution of Working Group III to the Fourth Assessment Report of the Intergovernmental Panel on Climate Change, Cambridge University Press, Cambridge United Kingdom and New York, USA.

Best, G.J.C. Earley, A. Faaij, U. Fritsche, A. Hester, S. Hunt, T. Iida, F. Johnson, P.M. Nastari, D.E. Newman, C.A. Opal, M. Otto, P. Read, R.E.H. Sims, S.C. Trindade, J. Bernard Tschirley and S. Zarilli, 2008. Sustainable Biofuels Consensus, Bellagio Declaration, prepared with support from the Rockefeller Foundation, April 2008, pp. 8.

Damen, K., 2001. Future prospects for biofuels in Brazil - a chain analysis comparison between production of ethanol from sugarcane and methanol from eucalyptus on short and longer term in Sao Paulo State-, Department of Science, Technology & Society - Utrecht University, November 2001, 70 pp.

De Vries, B.J.M., D.P. van Vuuren and M.M. Hoogwijk, 2007. Renewable energy sources: Their global potential for the first-half of the 21st century at a global level: An integrated approach. Energy Policy 35: 2590-2610.

Dornburg, V., A. Faaij, H. Langeveld, G. van de Ven, F. Wester, H. van Keulen, K. van Diepen, J. Ros, D. van Vuuren, G.J. van den Born, M. van Oorschot, F. Smout, H. Aiking, M. Londo, H. Mozaffarian, K. Smekens, M. Meeusen, M. Banse, E. Lysen and S. van Egmond, 2008. Biomass Assessment: Assessment of global biomass potentials and their links to food, water, biodiversity, energy demand and economy – Main Report, Study performed by Copernicus Institute – Utrecht University, MNP, LEI, WUR-PPS, ECN, IVM and the Utrecht Centre for Energy Research, within the framework of the Netherlands Research Programme on Scientific Assessment and Policy Analysis for Climate Change. Reportno: WAB 500102012, January 2008. 85 pp. + Appendices.

Faaij, A.P.C., 2006. Modern biomass conversion technologies. Mitigation and Adaptation Strategies for Global Change 11: 335-367.

FAO, 2008. State of Food and Agriculture - Biofuels: prospects, risks and opportunities, Food and Agirculture Organisation of the United Nations, Rome, Italy.

GBEP, 2008. A Review of the Current State of Bioenergy Development in G8 +5 Countries, Global Bioenergy Partnership, Rome, Italy.

Goldemberg, J., S. Teixeira Coelho, P. Mário Nastari and O. Lucon, 2004. Ethanol learning curve – the Brazilian experience, Biomass and Bioenergy 26: 301-304.

Hamelinck, C. and A. Faaij, 2006. Outlook for advanced biofuels. Energy Policy 34: 3268-3283.

Hamelinck, C.N., G. van Hooijdonk and A.P.C. Faaij, 2005. Future prospects for the production of ethanol from ligno-cellulosic biomass. Biomass & Bioenergy 28: 384-410.

Hettinga, W.G., H.M. Junginger, S.C. Dekker, M. Hoogwijk, A.J. McAloon and K.B. Hicks, 2008. Understanding the Reductions in U.S. Corn Ethanol Production Costs: An Experience Curve Approach, Energy Policy, in press.

Hoogwijk, M., A. Faaij, B. de Vries and W. Turkenburg, 2008. Global potential of biomass for energy from energy crops under four GHG emission scenarios Part B: the economic potential. Biomass & Bioenergy, in press.

Hoogwijk, M., A. Faaij, B. Eickhout, B. de Vries and W. Turkenburg, 2005. Potential of biomass energy out to 2100, for four IPCC SRES land-use scenarios, Biomass & Bioenergy 29: 225-257.

Hunt, S., J. Easterly, A. Faaij, C. Flavin, L. Freimuth, U. Fritsche, M. Laser, L. Lynd, J. Moreira, S. Pacca, J.L. Sawin, L. Sorkin, P. Stair, A. Szwarc and S. Trindade, 2007. Biofuels for Transport: Global Potential and Implications for Energy and Agriculture prepared by Worldwatch Institute, for the German Ministry of Food, Agriculture and Consumer Protection (BMELV) in coordination with the German Agency for Technical Cooperation (GTZ) and the German Agency of Renewable Resources (FNR), ISBN: 1844074226, Published by EarthScan/James & James, April 2007336 pp.

International Energy Agency, 2004. Biofuels for transport – an international perspective, Office of Energy Efficiency, Technology and R&D, OECD/IEA, Paris, France.

International Energy Agency, 2006. World Energy Outlook 2006, IEA Paris, France.

Junginger, M., T. Bolkesjø, D. Bradley, P. Dolzan, A. Faaij, J. Heinimö, B. Hektor, Ø. Leistad, E. Ling, M. Perry, E. Piacente, F. Rosillo-Calle, Y. Ryckmans, P.-P. Schouwenberg, B. Solberg, E. Trømborg, A. da Silva Walter and M. de Wit, 2008. Developments in international bioenergy trade. Biomass & Bioenergy 2: 717-729.

Kahn Ribeiro, S., S. Kobayashi, M. Beuthe, J. Gasca, D. Greene, D.S. Lee, Y. Muromachi, P.J. Newton, S. Plotkin, D. Sperling, R. Wit and P.J. Zhou, 2007: Transport and its infrastructure. In: B. Metz, O.R. Davidson, P.R. Bosch, R. Dave and L.A. Meyer (eds.), Climate Change 2007: Mitigation contribution of Working Group III to the Fourth Assessment Report of the Intergovernmental Panel on Climate Change, Cambridge University Press, Cambridge United Kingdom and New York, USA.

Kruithof, T., 2007. Drive the future, techno-economic comparison of fuel cell, serial hybrid and internal combustion engine drivechains for light duty vehicles. Copernicus Institute – Utrecht University, department of Science, Technology & Society. Report no.: NWS-I-2007-25, November 2007. 81 pp.

Kutas, G., 2008. Brazil's Sugar and Ethanol Industry: Endless Growth?. Presentation in Brussels, June 2008. Available at: www.unica.com.br.

Licht, F.O., 2007. World Bioethanol and Biofuels Report 5: 17.

Macedo, I.d.C., J.E.A. Seabra and J.E.A.R. Da Silva, 2008. Green house gases emissions in the production and use of ethanol from sugarcane in Brazil: The 2005/2006 averages and a prediction for 2020. Biomass & Bioenergy 32: 582-595.

Moreira, J.R., 2006. Global biomass energy potential. Mitigation and Adaptation Strategies for Global Change 11: 313-342.

OECD/FAO, 2007. Agricultural Outlook 2007-2016, Paris-France.

RFA, 2008. Changing the Climate - Ethanol Industry Outlook 2008. Renewable Fuels Association, page 16.

Rosillo-Calle, F. and L.A.B. Cortez, 1998. Towards Pro-Alcool II - a review of the Brazilian bioethanol programme, Biomass & Bioenergy 14: 115-124.

Smeets, E., M. Junginger, A. Faaij, A. Walter, P. Dolzan and W. Turkenburg, 2008. The sustainability of Brazilian Ethanol; an assessment of the possibilties of certified production. Biomass & Bioenergy 32: 781-813.

Valdes, C., 2007. Ethanol demand driving the expansion of Brazil's sugar industry. Sugar and Sweeteners Outlook/SSS-249: 31-38.

Van Dam, J., M. Junginger, A. Faaij, I. Jürgens, G. Best and U. Fritsche, 2008. Overview of recent developments in sustainable biomass certification. Biomass & Bioenergy 32: 749-780.

Wall-Bake, J.D., M. Junginger, A, Faaij, T. Poot and A. da Silva Walter, 2008. Explaining the experience curve: Cost reductions of Brazilian ethanol from sugarcane. Biomass & Bioenergy, in press.

Walter, A., F. Rosillo-Calle, P. Dolzan, E. Piacente and K. Borges da Cunha, 2008. Perspectives on fuel ethanol consumption and trade. Biomass & Bioenergy 32: 730-748.

WBCSD, 2004a. Mobility 2030: Meeting the Challenges to Sustainability. Available at: http://www.wbcsd.ch/

WBCSD, 2004b. IEA/SMP Model Documentation and Reference Projection. Fulton, L. and G. Eadsn (eds.). Available at: http://www.wbcsd.org/web/publications/mobility/smp-model-document.pdf.

Zarrilli. S., 2006. The emerging biofuels market: regulatory, trade and development implications. Geneva: United NationsConference on Trade and Development (UNCTAD); 2006. Available at: /www.unctad.org/en/docs/ditcted20064_en.pdfS.

Chapter 7
Biofuel conversion technologies

Andre Faaij

1. Introduction

In the current heated societal debate about the sustainability of biofuels, usually a distinction is made between so-called 'first' and 'second' generation biofuels. A large number of options to produce biomass from biofuel is used or are possible (a simplified overview of options is given in Figure 1). Although definitions differ between publications, first generation biofuels typically are produced from food crops as oilseeds (rapeseed, palm oil), starch crops (cereals, maize) or sugar crops (sugar beet and sugarcane). Conversion technologies are commercial and typically feedstock costs dominate the overall biofuel production costs. Furthermore,

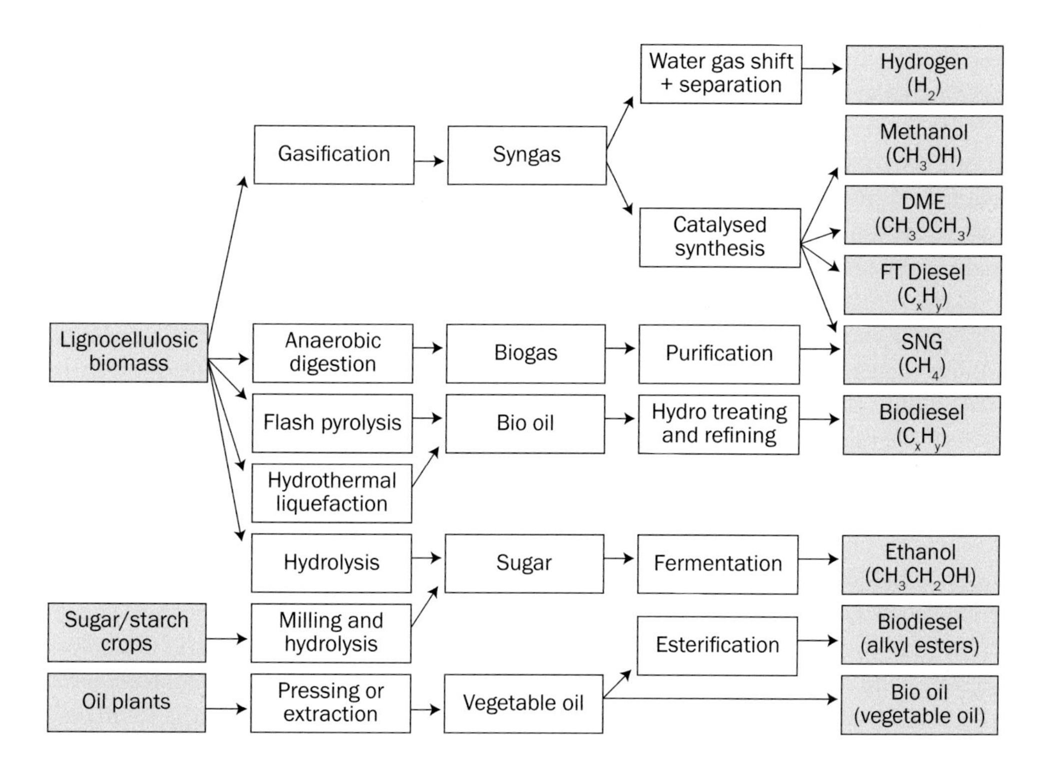

Figure 1. different existing and possible biofuel production routes (Hamelinck and Faaij, 2006). This is a simplified overview; other production chains are possible for example by combining conversion pathways, e.g. combined ethanol and biogas production, ethanol production and gasification of lignine for synfuels and integrated concepts with other industrial processes (pulp & paper plants) or bio refineries.

in particular when food crops are used grown in temperate climates (i.e. the US and the EU), costs are typically high due to high feedstock costs and the net overall avoided GHG emissions range between 20-50% compared to conventional gasoline or diesel (Fulton, 2004, Hunt *et al.*, 2007). Another constraint is that such food crops need to be produced on better quality land and increased demand directly competes with food markets. This has recently led to a wide range of estimates on the presumed impact of biofuel production on food prices (FAO, 2008), ranging between 3 up to 75%. However, sugarcane based ethanol production is a notable exception to these key concerns. Overall production costs as achieved in Brazil are competitive without subsidies, net GHG balance achieves 80-90% reduction and sugar prices have remained constant or have decreased slightly over the past years, despite strong increases in ethanol production from sugarcane.

Palm oil, in turn, although far less important as feedstock for biofuel production has been at the centre of the sustainability debate, because it's production is directly linked to loss of rainforest and peat lands in South-East Asia. Nevertheless, palm oil is an efficient and high yield crop to produce vegetal oil (Fulton, 2004). Recently, interest in Jatropha, a oil crop that can be grown in semi-arid conditions is growing, but commercial experience is very limited to date.

Second generation biofuels are not commercially produced at this stage, although in various countries demonstration projects are ongoing. 2nd generation biofuels are to be produced from lignocellulosic biomass. In lignocellulose, typically translated as biomass from woody crops or grasses and residue materials such as straw, sugars are chemically bound in chains and cannot be fermented by conventional micro-organisms used for production of ethanol from sugars and the type of sugars are different than from starch or sugar crops. In addition, woody biomass contains (variable) shares of lignine, that cannot be converted to sugars. Thus, more complex conversion technology is needed for ethanol production. Typical processes developed include advanced pre-treatment and enzymatic hydrolysis, to release individual sugars. Also fermentation of C5 instead of C6 sugars is required. The other key route being developed is gasification of lignocellulosic biomass, subsequent production of clean syngas that can be used to produce a range of synthetic biofuels, including methanol. DME and synthetic hydrocarbons (diesel). Because lignocellulosic biomass can origin from residue streams and organic wastes (that do in principle not lead to extra land-use when utilised), from trees and grasses that can also be grown on lower quality land (including degraded and marginal lands), it is thought that the overall potential of such routes is considerably larger on longer term than for 1st generation biofuels. Also, the inherently more extensive cultivation methods lead to very good net GHG balances (around 90% net avoided emissions) and ultimatly, they are thought to deliver competitive biofuels, due to lower feedstock costs, high overall chain efficiency, net energy yield per hectare, assuming large scale conversion.

This chapter gives an overview of the options to produce fuels from biomass, addressing current performance and the possible technologies and respective performance levels on longer term. It focuses on the main currently deployed routes to produce biofuels and on the key chains that are currently pursued for production of 2nd generation biofuels. Furthermore, an outlook on future biomass supplies is described in section 2, including a discussion of the impact of sustainability criteria and main determining factors and uncertainties. The chapter is finalized with a discussion of projections of the possible longer term role of biofuels on a global scale and the respective contribution of first and second generation biofuels.

2. Long term potential for biomass resources.

This section discusses a integral long term outlook on the potential global biomass resource base, including the recent sustainability debate and concerns. This assessment covered on global biomass potential estimates, focusing on the various factors affecting these potentials, such as food supplies, water use, biodiversity, energy demands and agro-economics (Dornburg *et al.*, 2008). The assessment focused on the relation between estimated biomass potentials and the availability and demand of water, the production and demand of food, the demand for energy and the influence on biodiversity and economic mechanisms.

The biomass potential, taken into account the various uncertainties as analysed in this study, consists of three main categories of biomass:

1. *Residues* from forestry and agriculture and organic waste, which in total represent between 40 - 170 EJ/yr, with a mean estimate of around 100 EJ/yr. This part of the potential biomass supplies is relatively certain, although competing applications may push the net availability for energy applications to the lower end of the range. The latter needs to be better understood, e.g. by means of improved models including economics of such applications.
2. *Surplus forestry*, i.e. apart from forestry residues an additional amount about 60-100 EJ/yr of surplus forest growth is likely to be available.
3. Biomass produced via *cropping systems*:
   a. A lower estimate for energy crop production *on possible surplus good quality agricultural and pasture lands*, including far reaching corrections for water scarcity, land degradation and new land claims for nature reserves represents an estimated 120 EJ/yr (‘*with exclusion of areas*’ in Figure 2).
   b. The potential contribution of *water scarce, marginal and degraded lands* for energy crop production, could amount up to an additional 70 EJ/yr. This would comprise a large area where water scarcity provides limitations and soil degradation is more severe and excludes current nature protection areas from biomass production (‘*no exclusion*’ in Figure 2).
   c. *Learning in agricultural technology* would add some 140 EJ/yr to the above mentioned potentials of energy cropping.

The three categories added together lead to a biomass supply potential of up to about 500 EJ.

Energy demand models calculating the amount of biomass used if energy demands are supplied cost-efficiently at different carbon tax regimes, estimate that in 2050 about 50-250 EJ/yr of biomass are used. At the same time, scenario analyses predict a global primary energy use of about 600 – 1040 EJ/yr in 2050 (the two right columns of Figure 2). Keep in mind that food demand of around 9 billion people in 2050 are basically met in those scenario's.

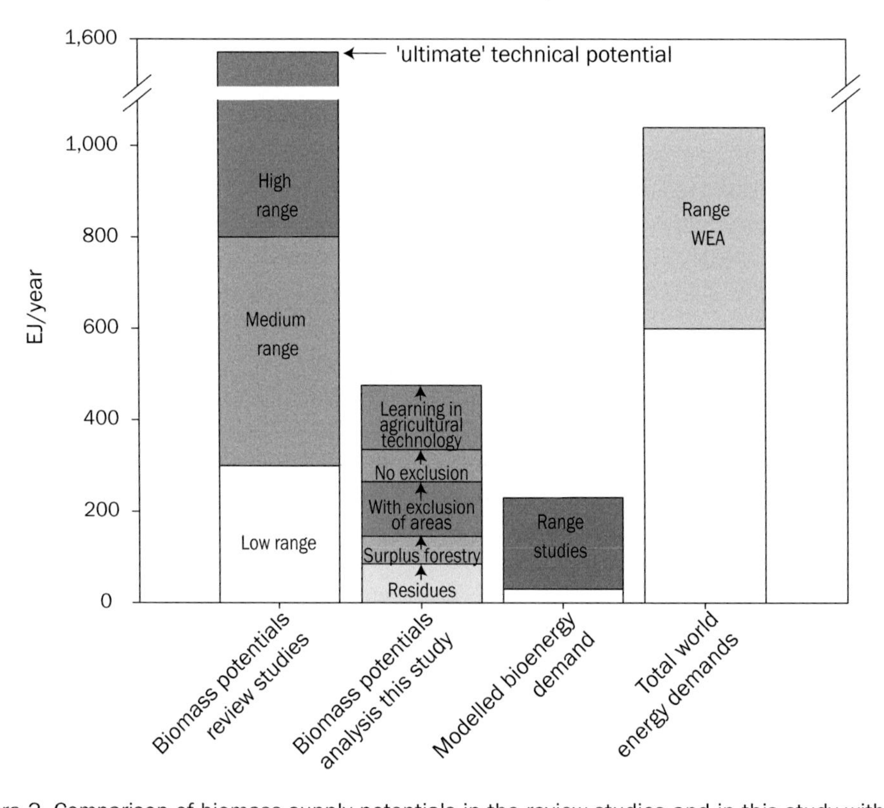

Figure 2. Comparison of biomass supply potentials in the review studies and in this study with the modelled demand for biomass and the total world energy demand, all for 2050 (Dornburg *et al.*, 2008). EJ = Exajoule (current global energy use amounts about 470 EJ at present). The first bar from the left represents the range of biomass energy potentials found in different studies, the second presents the results generated in (Dornburg *et al.*, 2008), taking a variety of sustainability criteria into account (such as water availability, biodiversity protection and soil quality), the third bar shows currently available estimates of biomass demand for energy from long term scenario studies and the fourth bar shows the range of projections of total global energy use in 2050.

In principle, biomass potentials are likely to be sufficient to allow biomass to play a significant role in the global energy supply system. Current understanding of the potential contribution of biomass to the future world energy supply is that the total technical biomass supplies could range from about 100 EJ using only residues up to an ultimate technical potential of 1500 EJ/yr potential per year. The medium range of estimates is between 300 and 800 EJ/yr (first column of Figure 2).

This study (Dornburg *et al.*, 2008) has confirmed that annual food crops may not be suited as a prime feedstock for bioenergy, both in size of potentials and in terms of meeting a wide array of sustainability criteria, even though annual crops can be a good alternative under certain circumstances. Perennial cropping systems, however, offer very different perspectives. These cannot only be grown on (surplus) agricultural and pasture lands, but also on more marginal and degraded lands, be it with lower productivity. At this stage there is still limited (commercial) experience with such systems for energy production, especially considering the more marginal and degraded lands and much more development, demonstration (supported by research) is needed to develop feasible and sustainable systems suited for very different settings around the globe. This is a prime priority for agricultural policy.

As summarized, the size of the biomass resource potentials and subsequent degree of utilisation depend on numerous factors. Part of those factors are (largely) beyond policy control. Examples are population growth and food demand. Factors that can be more strongly influenced by policy are development and commercialization of key technologies (e.g. conversion technology that makes production of fuels from lignocellulosic biomass and perennial cropping systems more competitive), e.g. by means of targeted RD&D strategies. Other areas are:

- Sustainability criteria, as currently defined by various governments and market parties.
- Regimes for trade of biomass and biofuels and adoption of sustainability criteria (typically to be addressed in the international arena, for example via the WTO).
- Infrastructure; investments in infrastructure (agriculture, transport and conversion) is still an important factor in further deployment of bioenergy.
- Modernization of agriculture; in particular in Europe, the Common Agricultural Policy and related subsidy instruments allow for targeted developments of both conventional agriculture and second generation bioenergy production. Such sustainable developments are however crucial for many developing countries and are a matter for national governments, international collaboration and various UN bodies.
- Nature conservation; policies and targets for biodiversity protection do determine to what extent nature reserves are protected and expanded and set standards for management of other lands.
- Regeneration of degraded lands (and required preconditions), is generally not attractive for market parties and requires government policies to be realized.

Current insights provide clear leads for further steps for doing so. In the criteria framework as defined currently by several governments, in roundtables and by NGO's, it is highlighted that a number of important criteria require further research and design of indicators and verification procedures. This is in particular the case for to the so-called 'macro-themes' (land-use change, biodiversity, macro-economic impacts) and some of the more complex environmental issues (such as water use and soil quality). Sustainability of biofuels and ongoing development around defining criteria and deployment of certification is discussed in Chapter 5 of this book by Neves do Amaral.

3. Technological developments in biofuel production

The previous section highlights the importance of lignocellulosic resources for achieving good environmental performance and reducing the risks of competition for land and with food production. This implies that different technologies are required to produce liquid fuels, compared to the currently dominant use of annual crops as maize and rapeseed. Sugarcane is however a notable exception given it's very high productivity, low production costs and good energy and GHG balance (Macedo *et al.*, 2004; Smeets *et al.*, 2008).

Three main routes can be distinguished to produce transportation fuels from biomass: gasification can be used to produce syngas from lignocellulosic biomass that can be converted to methanol, Fischer-Tropsch liquids, DiMethylEther (DME) and hydrogen. Production of ethanol can take place via direct fermentation of sugar and starch rich biomass, the most utilized route for production of biofuels to date, or this can be preceded by hydrolysis processes to convert lignocellulosic biomass to sugars first. Finally, biofuels can be produced via extraction from oil seeds (vegetal oil from e.g. rapeseed or palm oil), which can be esterified to produce biodiesel.

Other conversion routes and fuels are possible (such as production of butanol from sugar or starch crops) and production of biogas via fermentation. The above mentioned routes have however so far received most attention in studies and Research and Demonstration efforts.

3.1. Methanol, hydrogen and hydrocarbons via gasification

Methanol (MeOH), hydrogen (H_2) and Fischer Tropsch synthetic hydrocarbons (especially diesel), DME (DiMethylEther) and SNG (Synthetic Natural Gas) can be produced from biomass via gasification. All routes need very clean syngas before the secondary energy carrier is produced via relatively conventional gas processing methods. Here, focus lays on the first three fuels mentioned.

Several routes involving conventional, commercial, or advanced technologies under development, are possible. Figure 3 pictures a generic conversion flowsheet for this category of processes. A train of processes to convert biomass to required gas specifications precedes the methanol or FT reactor, or hydrogen separation. The gasifier produces syngas, a mixture of CO and H_2, and a few other compounds. The syngas then undergoes a series of chemical reactions. The equipment downstream of the gasifier for conversion to H_2, methanol or FT diesel is the same as that used to make these products from natural gas, except for the gas cleaning train. A gas turbine or boiler, and a steam turbine optionally employ the unconverted gas fractions for electricity co-production (Hamelinck *et al.*, 2004).

So far, commercial biofuels production via gasification does not take place, but interest is on the rise and development and demonstration efforts are ongoing in several OECD countries.

Overall energetic efficiencies of relatively 'conventional' production facilities, could be close to 60% (on a scale of about 400 MW_{th} input). Deployment on large scale (e.g over 1000 MW_{th}) is required to benefit maximally from economies of scale, which are inherent to this type of installations. Such capacities are typical for coal gasification. The use of coal gasifiers and feeding of pre-treated biomass (e.g. via torrefaction or pyrolysis oils) could prove one of the shorter term options to produce 2nd generation biofuels efficiently. This conversion route has a strong position from both efficiency and economic perspective (Hamelinck *et al.*, 2004; Hamelinck and Faaij, 2002; Tijmensen et al, 2002; Williams *et al.*, 1995). Generic performance ranges resulting from various pre-engineering studies are reported in Figure 3.

The findings of the previously published papers can be summarised as follows: gasification-based fuel production systems that apply pressurised gasifiers have higher joint fuel and electricity energy conversion efficiencies than atmospheric gasifier-based systems. The total efficiency is also higher for once-through configurations, than for recycling configurations that aim at maximising fuel output. This effect is strongest for FT production, where (costly) syngas recycling not only introduces temperature and pressure leaps, but also 'material leaps' by reforming part of the product back to syngas. For methanol and hydrogen, however,

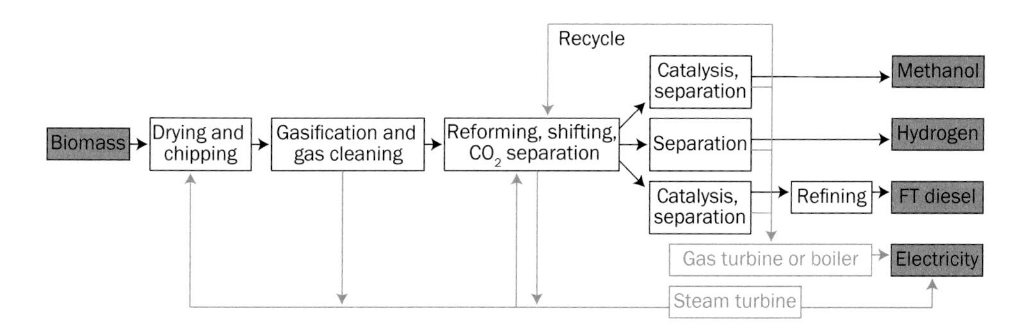

Figure 3. Generic process scheme for production of synthetic biofuels via gasification (Hamelinck and Faaij, 2006).

maximised fuel production, with little or no electricity co-production, generally performs economically somewhat better than once-through concepts.

Hot (dry) gas cleaning generally improves the total efficiency, but the economical effects are ambivalent, since the investments also increase. Similarly, CO_2 removal does increase the total efficiency (and in the FT reaction also the selectivity), but due to the accompanying increase in investment costs this does not decrease the product costs. The bulk of the capital investment is in the gasification and oxygen production system, syngas processing and power generation units. These parts of the investment especially profit from cost reductions at larger scales. Also, combinations with enriched air gasification (eliminating the expensive oxygen production assumed for some methanol and hydrogen concepts) may reduce costs further.

Several technologies considered here are not yet fully proven or commercially available. Pressurised (oxygen) gasifiers still need further development. At present, only a few pressurised gasifiers, operating at relatively small scale, have proved to be reliable. Consequently, the reliability of cost data for large-scale gasifiers is uncertain. A very critical step in all thermal systems is gas cleaning. It still has to be proven whether the (hot) gas cleaning section is able to meet the strict cleaning requirements for reforming, shift and synthesis. Liquid phase reactors (methanol and FT) are likely to have better economies of scale. The development of ceramic membrane technology is crucial to reach the projected hydrogen cost level. For FT diesel production, high CO conversion, either once through or after recycle of unconverted gas, and high C5+ selectivity are important for high overall energy efficiencies. Several units may be realised with higher efficiencies than considered in this paper: new catalysts and carrier liquids could improve liquid phase methanol single pass efficiency. At larger scales, conversion and power systems (especially the combined cycle) have higher efficiencies, further stressing the importance of achieving economies of scale for such concepts.

3.2. Production of ethanol from sugarcane

Ethanol production from sugarcane has established a strong position in Brazil and increasingly in other countries in tropical regions (such as India, China and various countries in Sub-Saharan Africa). Production costs of ethanol in Brazil have steadily declined over the past few decades and have reached a point where ethanol is competitive with production costs of gasoline (Wall-Bake *et al.*, 2008). As a result, bioethanol is no longer financially supported in Brazil and competes openly with gasoline.

Large scale production facilities, better use of bagasse and trash residues from sugarcane production e.g. with advanced (gasification based) power generation or hydrolysis techniques (see below) and further improvements in cropping systems, offer further perspectives for sugarcane based ethanol production.

Improvement options for sugarcane based ethanol production are plentiful (Damen, 2001; Groen, 1999). It is expected that the historic cost decreases and productivity increments will continue. An analysis of historic and potential future improvements in economic performance of ethanol production in Brazil (Wall Bake *et al.*, 2008) concludes that if improvements in sugarcane yield, logistics (e.g. green can harvesting techniques and utilisation of sugarcane trash), overall efficiency improvement in the sugar mills and ethanol production (e.g. by full electrification and advanced distillation technology) as well as the use of hydrolysis technology for conversion of bagasse and trash to ethanol, ethanol yields per hectare of land may even be tripled compared to current average production.

The key limitations for sugarcane production are climatic and the required availability of good quality soils with sufficient and the right rainfall patterns.

3.3. Ethanol from (ligno)-cellulosic biomass

Hydrolysis of cellulosic (e.g. straw) and lignocellulosic (woody) biomass can open the way towards low cost and efficient production of ethanol from these abundant types of biomass. The conversion is more difficult than for sugar and starch because from lignocellulosic materials, first sugars need to be produced via hydrolysis. Lignocellulosic biomass requires pretreatment by mechanical and physical actions (e.g. steam) to clean and size the biomass, and destroy its cell structure to make it more accessible to further chemical or biological treatment. Also, the lignin part of the biomass is removed, and the hemicellulose is hydrolysed (saccharified) to monomeric and oligomeric sugars. The cellulose can then be hydrolysed to glucose. Also C5 sugars are formed, which require different yeasts to be converted to ethanol. The sugars are fermented to ethanol, which is to be purified and dehydrated. Two pathways are possible towards future processes: a continuing consolidation of hydrolysis-fermentation reactions in fewer reactor vessels and with fewer micro organisms, or an optimisation of separate reactions. As only the cellulose and hemicellulose can be used in the process, the lignin is used for power production (Figure 4).

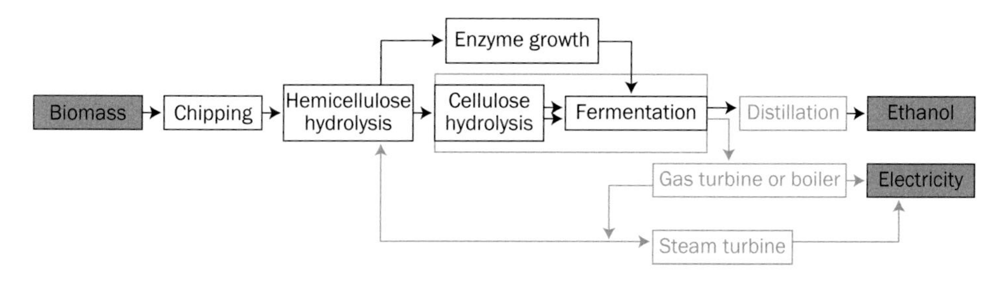

Figure 4. Generic process scheme for the production of ethanol from lignocellulosic biomass.

To date, acid treatment is an available process, which is so far relatively expensive and inefficient. Enzymatic treatment is commercially unproven but various test facilities have been built in North America and Sweden. The development of various hydrolysis techniques has gained major attention over the past 10 years or so, particularly in Sweden and the United States. Because breakthroughs seem to be necessary on a rather fundamental level, it is relatively uncertain how fast attractive performance levels can be achieved (Hamelinck *et al.*, 2005).

Assuming, however, that mentioned issues are resolved and ethanol production is combined with efficient electricity production from unconverted wood fractions (lignine in particular), ethanol costs could come close to current gasoline prices (Lynd *et al.*, 2005): as low as 12 Euroct/litre assuming biomass costs of about 2 Euro/GJ. Overall system efficiencies (fuel + power output) could go up to about 70% (LHV).

It should be noted though that the assumed conversion extent of (hemi)cellulose to ethanol by hydrolysis fermentation is close to the stoichiometric maximum. There is only little residual material (mainly lignin), while the steam demand for the chosen concepts is high. This makes the application of BIG/CC unattractive at 400MWHHV. Developments of pre-treatment methods and the gradual ongoing reactor integration are independent trends and it is plausible that at least some of the improved performance will be realised in the medium-term. The projected long-term performance depends on development of technologies that have not yet passed laboratory stage, and that may be commercially available earlier or later than 20 years from now. This would mean either a more attractive ethanol product cost in the medium-term, or a less attractive cost in the long-term. The investment costs for advanced hemicellulose hydrolysis methods is still uncertain. Continuing development of new micro-organisms is required to ensure fermentation of xylose and arabinose, and decrease the cellulase enzyme costs.

The hydrolysis technology can also boost the competitiveness of existing production facilities (e.g. by converting available crop and process residues), which provides an important market niche on short term.

Table 1. gives an overview of estimates for costs of various fuels that can be produced from biomass (Faaij, 2006). A distinction is made between performance levels on the short and on the longer term. Generally spoken, the economy of 'traditional' fuels like Rapeseed MethylEsther and ethanol from starch and sugar crops in moderate climate zones is poor at present and unlikely to reach competitive price levels in the longer term. Also, the environmental impacts of growing annual crops are not as good as perennials because per unit of product considerable higher inputs of fertilizers and agrochemicals are needed. In addition, annual crops on average need better quality land than perennials to achieve good productivities.

Production of methanol (and DME), hydrogen, Fischer-Tropsch liquids and ethanol produced from lignocellulosic biomass that offer good perspectives and competitive fuel prices in the longer term (e.g. around 2020). Partly, this is because of the inherent lower feedstock prices and versatility of producing lignocellulosic biomass under varying circumstances. Section 2 highlighted that a combination of biomass residues and perennial cropping systems on both marginal and better quality lands could supply a few hundred EJ by mid-century in a competitive cost range between 1-2 Euro/GJ (see also Hoogwijk *et al.*, 2005, 2008). Furthermore, as discussed in this paper, the (advanced) gasification and hydrolysis technologies under development have the inherent improvement potential for efficient and competitive production of fuels (sometimes combined with co-production of electricity).

Inherent to the advanced conversion concepts, it is relatively easy to capture (and subsequently store) a significant part of the CO_2 produced during conversion at relatively low additional costs. This is possible for ethanol production (where partially pure CO_2 is produced) and for gasification concepts. Production of syngas (both for power generation and for fuels) in general allows for CO_2 removal prior to further conversion. For FT production about half of the carbon in the original feedstock (coal, biomass) can be captured prior to the conversion of syngas to FT-fuels. This possibility allows for carbon neutral fuel production when mixtures of fossil fuels and biomass are used and negative emissions when biomass is the dominant or sole feedstock. Flexible new conversion capacity will allow for multiple feedstock and multiple output facilities, which can simultaneously achieve low, zero or even negative carbon emissions. Such flexibility may prove to be essential in a complex transition phase of shifting from large scale fossil fuel use to a major share of renewables and in particular biomass.

At the moment major efforts are ongoing to demonstrate various technology concepts discussed above. Especially in the US (but also in Europe), a number of large demonstration efforts is ongoing on production of ethanol from lignocellulosic biomass. IOGEN, a Canadian company working on enzymatic hydrolysis reported the production of 100,000 litres of ethanol from agricultural residues in September 2008. Also companies in India, China and Japan are investing substantially in this technology area.

Gasification for production of synfuels gets support in the US and more heavily in the EU. The development trajectory of the German company CHOREN (focusing on dedicated biomass gasification systems for production of FT liquids) is ongoing and stands in the international spotlights. Finland and Sweden have substantial development efforts ongoing, partly aiming for integration gasification technology for synfuels in the paper & pulp industry. Furthermore, co-gasification of biomass in (existing) coal gasifiers is an important possibility. This has for example been demonstrated in the Buggenum coal gasifiier in the Netherlands and currently production of synfuels is targeted.

Table 1. Performance levels for different biofuels production routes (Faaij, 2006).

Concept	Energy efficiency (HHV) + energy inputs	
	Short term	Long term
Hydrogen: via biomass gasification and subsequent syngas processing. Combined fuel and power production possible; for production of liquid hydrogen additional electricity use should be taken into account.	60% (fuel only) (+ 0.19 GJe/GJ H2 for liquid hydrogen)	55% (fuel) 6% (power) (+ 0.19 GJe/GJ H2 for liquid hydrogen)
Methanol: via biomass gasification and subsequent syngas processing. Combined fuel and power production possible	55% (fuel only)	48% (fuel) 12% (power)
Fischer-Tropsch liquids: via biomass gasification and subsequent syngas processing. Combined fuel and power production possible	45% (fuel only)	45% (fuel) 10% (power
Ethanol from wood: production takes place via hydrolysis techniques and subsequent fermentation and includes integrated electricity production of unprocessed components.	46% (fuel) 4% (power)	53% (fuel) 8% (power)
Ethanol from beet sugar: production via fermentation; some additional energy inputs are needed for distillation.	43% (fuel only) 0.065 GJe + 0.24 GJth/GJ EtOH	43% (fuel only) 0.035 GJe + 0.18 GJth/GJ EtOH
Ethanol from sugarcane: production via cane crushing and fermentation and power generation from the bagasse. Mill size, advanced power generation and optimised energy efficiency and distillation can reduce costs further on longer term.	85 litre EtOH per tonne of wet cane, generally energy neutral with respect to power and heat	95 litre EtOH per tonne of wet cane. Electricity surpluses depend on plant lay-out and power generation technology.
Biodiesel RME: takes places via extraction (pressing) and subsequent esterification. Methanol is an energy input. For the total system it is assumed that surpluses of straw are used for power production.	88%; 0.01 GJe + 0.04 GJ MeOH per GJ output Efficiency power generation on shorter term: 45%, on longer term: 55%	

Assumed biomass price of clean wood: 2 Euro/GJ. RME cost figures varied from 20 Euro/GJ (short term) to 12 Euro/GJ (longer term), for sugar beet a range of 12 to 8 Euro/GJ is assumed. All figures exclude distribution of the fuels to fueling stations.

For equipment costs, an interest rate of 10%, economic lifetime of 15 years is assumed. Capacities of conversion unit are normalized on 400 MWth input on shorter term and 1000 MWth input on longer term.

Investment costs (Euro/kWth input capacity)		O&M (% of inv.)	Estimated production costs (Euro/GJ fuel)	
Short term	**Long term**		**Shorter term**	**Longer term**
480 (+ 48 for liquefying)	360 (+ 33 for liquefying)	4	9-12	4-8
690	530	4	10-15	6-8
720	540	4	12-17	7-9
350	180	6	12-17	5-7
290	170	5	25-35	20-30
100 (wide range applied depending on scale and technology applied)	230 (higher costs due to more advanced equipment)	2	8-12	7-8
150 (+ 450 for power generation from straw)	110 (+ 250 for power generation from straw)	5 4	25-40	20-30

Industrial interest in those areas comes from the energy sector, biotechnology as well as chemical industry. Given the policy targets on (second generation) biofuels in North America and the EU, high oil prices and increased pressure to secure sustainable production of biofuels (e.g. avoiding conflicts with food production and achieve high reduction in GHG emissions), pressure on both the market and policy to commercialize those technologies is high. When turn-key processes are available is still uncertain, but such breakthroughs may be possible already around 2010.

4. Energy and greenhouse gas balances of biofuels

4.1. Energy yields

The energy yield per unit of land surfaces resources depends to a large extent on the crop choice and the efficiency of the entire energy conversion route from 'crop to drop'. This is illustrated by the figures in Table 2. It is important to stress that when lignocellulose is the feedstock of choice production is not constrained to arable land, but amounts to the sum of residues and production from degraded/marginal lands not used for current food production. Ultimately, this will be the preferred option in most cases.

Table 2. Indicative ranges for biomass yield and subsequent fuel production per hectare per year for different cropping systems in different settings. Starch and sugar crops assume conversion via fermentation to ethanol and oil crops to biodiesel via esterification (commercial technology at present). The woody and grass crops require either hydrolysis technology followed by ethanol or gasification to syngas to produce synthetic fuel (both not yet commercial conversion routes).

Crop	Biomass yield (odt/ha/yr)	Energy yield in fuel (GJ/ha/yr)
Wheat	4-5	~50
Maize	5-6	~60
Sugar beet	9-10	~110
Soy bean	1-2	~20
Sugarcane	5-20	~180
Palm oil	10-15	~160
Jathropha	5-6	~60
SRC temperate climate	10-15	100-180
SRC tropical climate	15-30	170-350
Energy grasses good conditions	10-20	170-230
Perennials marginal/degraded lands	3-10	30-120

4.2. Greenhouse gas balances

The net emissions over the full life cycle of biofuels – from changes in land use to combustion of fuels – that determine their impact on the climate. Research on net emissions is far from conclusive, and estimates vary widely. Calculations of net GHG emissions are highly sensitive to assumptions about system boundaries and key parameter values – for example, land use changes and their impacts, which inputs are included, such as energy embedded in agricultural machinery and how various factors are weighted.

The primary reasons for differing results are different assumptions made about cultivation, and conversion or valuation of co-products. (Larson, 2005), who reviewed multiple studies, found that the greatest variations in results arose from the allocation method chosen for co-products, and assumptions about N_2O emissions and soil carbon dynamics. In addition, GHG savings will vary from place to place – according to existing incentives for GHG reductions, for example. And the advantages of a few biofuels (e.g. sugarcane ethanol in Brazil) are location specific. As a result, it is difficult to compare across studies; however, despite these challenges, some of the more important studies point to several useful conclusions.

This analysis notwithstanding, the vast majority of studies have found that, even when all fossil fuel inputs throughout the life cycle are accounted for, producing and using biofuels made from current feedstock result in substantial reductions in GHG emissions relative to petroleum fuels.

In general, of all potential feedstock options, producing ethanol from maize results in the smallest decrease in overall emissions. The greatest benefit, meanwhile, comes from ethanol produced from sugarcane grown in Brazil (or from using cellulose or wood waste as feedstock). Several studies have assessed the net emissions reductions resulting from sugarcane ethanol in Brazil, and all have concluded that the benefits far exceed those from grain-based ethanol produced in Europe and the United States.

Fulton (2004) attributes the lower life-cycle climate impacts of Brazilian sugarcane ethanol to two major factors: First, cane yields are high and require relatively low inputs of fertilizer, since Brazil has better solar resources and high soil productivity. Second, almost all conversion plants use bagasse (the residue that remains after pressing the sugar juice from the cane stalk) for energy, and many recent plants use co-generation (heat and electricity), enabling them to feed electricity into the grid. As such, net fossil energy requirements are near zero, and in some cases could be below zero. (In addition, less energy is required for processing because there is no need for the extra step to break down starch into simple sugars. Because most process energy in Brazil is already renewable, this does not really play a role.)

According to Larson (2005), conventional grain- and oilseed-based biofuels can offer only modest reductions in GHG emissions. The primary reason for this is that they represent only a small portion of the above ground biomass. He estimates that, very broadly, biofuels from grains or seeds have the potential for a 20–30 percent reduction in GHG emissions per vehicle-kilometer, sugar beets can achieve reductions of 40–50 percent, and sugarcane (average in southeast Brazil) can achieve a reduction of 90 percent.

Other new technologies under development also offer the potential to dramatically increase yields per unit of land and fossil input, and further reduce life-cycle emissions. The cellulosic conversion processes for ethanol offers the greatest potential for reductions because feedstock can come from the waste of other products or from energy crops, and the remaining parts of the plant can be used for process energy.

Larson (2005) projects that future advanced cellulosic processes (to ethanol, F-T diesel, or DME) from perennial crops could bring reductions of 80–90 percent and higher. According to Fulton *et al.* (2004), net GHG emissions reductions can even exceed 100 percent if the feedstock takes up more CO_2 while it is growing than the CO_2-equivalent emissions released during its full life cycle (for example, if some of it is used as process energy to offset coal-fired power).

Typical estimates for reductions from cellulosic ethanol (most of which comes from engineering studies, as few large-scale production facilities exist to date) range from 70–90 percent relative to conventional gasoline, according to Fulton (2004), though the full range of estimates is far broader.

Figure 5 shows the range of estimated possible reductions in emissions from wastes and other next-generation feedstock relative to those from current-generation feedstock and technologies.

4.3. Chain efficiency of biofuels

When the use of such 'advanced' biofuels (especially hydrogen and methanol) in advanced hybrid or Fuel Cell Vehicles (FCV's) is considered, the overall chain ('tree - to – tyre') efficiency can drastically improve compared to current bio-diesel or maize or cereal derived ethanol powered Internal Combustion Engine Vehicles; the effective number of kilometres that can be driven per hectare of energy crops could go up with a factor of 5 (from a typical current 20,000 km/ha for a middle class vehicle run with RME up to over 100,000 km/ha for advanced ethanol in an advanced hybrid or FCV (Hamelinck and Faaij, 2002)). Note though, that the current exception to this performance is sugarcane based ethanol production; in Brazil the better plantations yield some 8,000 litre ethanol/ha*yr, or some 70,000 km/yr for a middle class vehicle at present. In the future, those figures can improve

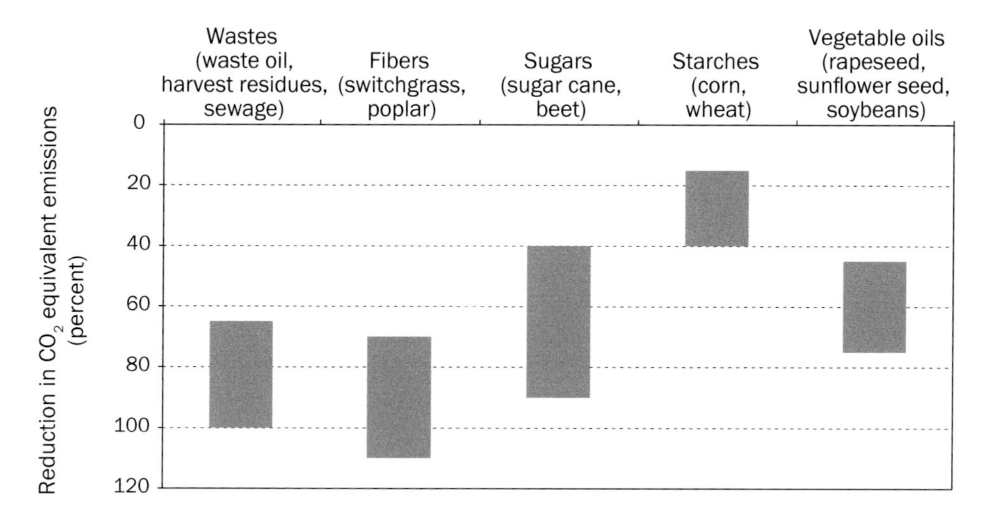

Figure 5. Reductions in greenhouse gas emissions per vehicle-kilometre, by feedstock and associated refining technology (taken from Fulton, 2004).

further due to better cane varieties, crop management and efficiency improvement in the ethanol production facilities (Damen, 2001).

Furthermore, FCV's (and to a somewhat lesser extent advanced hybrids) offer the additional and important benefits of zero or near zero emission of compounds like NO_x, CO, sulphur dioxide, hydrocarbons and small dust particulates, which are to a large extent responsible for poor air quality in many urban zones in the world. Table 3 provides a quantification of the range of kilometres that can be driven with different biofuel-vehicle combinations expressed per hectare. The ranges are caused by different yield levels for different land-types and variability and uncertainties in conversion and vehicle efficiencies. However, overall, there are profound differences between first and second generation biofuels I favour of the latter.

4.4. Future expectations on biofuels

The future biofuels and specifically the bioethanol market is uncertain. There are fundamental drivers (climate, oil prices and availability, rural development) that push for further development of biofuels. On the one hand, recent developments and public debate point towards conflicts with land use, food markets, poor GHG performance (especially when indirect land-use changes are assumed caused by biofuel production) and, even with high oil prices, high levels of subsidy for biofuels in e.g. Europe and the United States. Recently, policy targets (as discussed in chapter 5 of this book) set for biofuels are rediscussed in the EU, as well as in China. In most key markets (EU, US, China), the role of biofuels is increasingly connected to rapid deployment of 2nd generation technologies. The bulk of the growth beyond 2015 or so should be realized via such routes.

Table 3. Distance that can be driven per hectare of feedstock for several combinations of fuels and engines, derived from the net energy yield and vehicle efficiency as reported in (Hamelinck and Faaij, 2006). ICEV = Internal Combustion Engine Vehicle, FCV = Fuel Cell Vehicle.

Feedstock	Fuel	Engine	Distance (thousands km/ha)	
			Short term	Long term
Lignocellulose	Hydrogen	ICEV	26-37	80-97
		FCV	44-140	189-321
	Methanol	ICEV	34-49	75-287
		FCV	68-83	113-252
	FT	ICEV	22-38	56-167
		FCV	50-67	97-211
	Ethanol	ICEV	29-30	82-238
		FCV	38-72	129-240
Sugar beet	Ethanol	ICEV	15-37	57-88
		FCV	19-93	58-138
Rapeseed	RME	ICEV	5-28	15-79
		FCV	6-84	19-137

Some projections as published by the International Energy Agency (World Energy Outlook) and the OECD (Agricultural Outlook) focus on first generation biofuels only (even for projections to 2030 in the IEA-WEO). Biofuels meet 2.7% of world road-transport fuel demand by the end of the projection period in the Reference Scenario, up from 1% today. In the Alternative Scenario, the share reaches 4.6%, thanks to higher demand for biofuels but lower demand for road-transport fuels in total. The share remains highest in Brazil, though the pace of market penetration will be fastest in the European Union in both scenarios. The contribution of liquid biofuels to transport energy, and even more so to global energy supply, will remain limited. By 2030, liquid biofuels are projected to still supply only 3.0-3.5 percent of global transport energy demand. This is however also due to the key assumption that 2nd generation biofuel technology is not expected to become available to the market (IEA, 2006).

In the Agricultural Outlook, similar reasoning is followed be it for a shorter time frame (up to the year 2016), focusing on 1st generation biofuels. The outlook focuses in this respect on the implications of biofuel production on demand for food crops. In general, a slowdown in growth is expected (OECD, 2007).

Projections that take explicitly 2nd generation options into account are more rare, but studies that do so, come to rather different outlooks, especially in the timeframe exceeding 2020.

The IPCC, providing an assessment of studies that deal with both supply and demand of biomass and bioenergy. It is highlighted that biomass demand could lay between 70 – 130 EJ in total, subdivided between 28-43 EJ biomass input for electricity and 45-85 EJ for biofuels (Barker and Bashmakov, 2007; Kahn Ribeiro et al., 2007). Heat and biomass demand for industry are excluded in these reviews. It should also be noted that around that timeframe biomass use for electricity has become a less attractive mitigation option due to the increased competitiveness of other renewables (e.g. wind energy) and e.g. [and storage. At the same time, carbon intensity of conventional fossil transport fuels increases due to the increased use lower quality oils, tar sands and coal gasification.

In De Vries *et al.* (2007; based on the analyses of Hoogwijk *et al.* (2005, 2008)), it is indicated that the biofuel production potential around 2050 could lay between about 70 and 300 EJ fuel production capacity depending strongly on the development scenario. Around that time, biofuel production costs would largely fall in the range up to 15 U$/GJ, competitive with equivalent oil prices around 50-60 U$/barrel. This is confirmed by other by the information compiled in this chapter: it was concluded that the, sustainable, biomass resource base, without conflicting with food supplies, nature preservation and water use, could indeed be developed to a level of over 300 EJ in the first half of this century.

5. Final remarks

Biomass cannot realistically cover the whole world's future energy demand. On the other hand, the versatility of biomass with the diverse portfolio of conversion options, makes it possible to meet the demand for secondary energy carriers, as well as bio-materials. Currently, production of heat and electricity still dominate biomass use for energy. The question is therefore what the most relevant future market for biomass may be.

For avoiding CO_2 emissions, replacing coal is at present a very effective way of using biomass. For example, co-firing biomass in coal-fired power stations has a higher avoided emission per unit of biomass than when displacing diesel or gasoline with ethanol or biodiesel. However, replacing natural gas for power generation by biomass, results in levels of CO_2 mitigation similar to second generation biofuels. Net avoided GHG emissions therefore depend on the reference system and the efficiency of the biomass production and utilisation chain. In the future, using biomass for transport fuels will gradually become more attractive from a CO_2 mitigation perspective because of the lower GHG emissions for producing second generation biofuels and because electricity production on average is expected to become less carbon-intensive due to increased use of wind energy, PV and other solar-based power generation, carbon capture and storage technology, nuclear energy and fuel shift from coal to natural gas. In the shorter term however, careful strategies and policies are needed to avoid brisk allocation of biomass resources away from efficient and effective utilisation in power and heat production or in other markets, e.g. food. How this is to be done optimally will differ from country to country.

First generation biofuels in temperate regions (EU, North America) do not offer a sustainable possibility in the long term: they remain expensive compared to gasoline and diesel (even at high oil prices), are often inefficient in terms of net energy and GHG gains and have a less desirable environmental impact. Furthermore, they can only be produced on higher quality farmland in direct competition with food production. Sugarcane based ethanol production and to a certain extent palm oil and Jatropha oilseeds are notable exceptions to this given their high production efficiencies and low(er) costs.

Especially promising are the production via advanced conversion concepts biomass-derived fuels such as methanol, hydrogen, and ethanol from lignocellulosic biomass. Ethanol produced from sugarcane is already a competitive biofuel in tropical regions and further improvements are possible. Both hydrolysis-based ethanol production and production of synfuels via advanced gasification from biomass of around 2 Euro/GJ can deliver high quality fuels at a competitive price with oil down to US$55/ barrel. Net energy yields for unit of land surface are high and up to a 90% reduction in GHG emissions can be achieved. This requires a development and commercialization pathway of 10-20 years, depending very much on targeted and stable policy support and frameworks.

However, commercial deployment of these technologies does not have to be postponed for such time periods. The two key technological concepts that have shorter term opportunities (that could be seen as niches) for commercialization are:

1. Ethanol: 2nd generation can build on the 1st generation infrastructure by being built as 'add-ons' to existing factories for utilisation of crop residues. One of the best examples is the use of bagasse and trash at sugar mills that could strongly increase the ethanol output from sugarcane
2. Synfuels via gasification of biomass: can be combined with coal gasification as currently deployed for producing synfuels (such as DME, Fischer-Tropsch and Methanol) to obtain economies of scale and fuel flexibility. Carbon capture and storage can easily be deployed with minimal additional costs and energy penalties as an add-on technology.

The biomass resource base can become large enough to supply 1/3 of the total world's energy needs during this century. Although the actual role of bioenergy will depend on its competitiveness with fossil fuels and on agricultural policies worldwide, it seems realistic to expect that the current contribution of bioenergy of 40-55 EJ per year will increase considerably. A range from 200 to 400 EJ may be observed looking well into this century, making biomass a more important energy supply option than mineral oil today. Considering lignocellulosic biomass, about half of the supplies could originate from residues and biomass production from marginal/degrade lands. The other half could be produced on good quality agricultural and pasture lands without jeopardizing the worlds food supply, forests and biodiversity. The key pre-condition to achieve this goal is increased agricultural land-use efficiency, including livestock production, especially in developing regions. Improvement potentials of agriculture and livestock are substantial, but exploiting such potentials is a challenge.

References

Barker, T. and I. Bashmakov, 2007. Mitigation from a cross-sectoral perspective, In: B. Metz, O.R. Davidson, P.R. Bosch, R. Dave and L.A. Meyer (eds.), Climate Change 2007: Mitigation contribution of Working Group III to the Fourth Assessment Report of the Intergovernmental Panel on Climate Change, Cambridge University Press, Cambridge United Kingdom and New York, USA.

Damen, K., 2001. Future prospects for biofuels in Brazil - a chain analysis comparison between production of ethanol from sugarcane and methanol from eucalyptus on short and longer term in Sao Paulo State-, Department of Science, Technology & Society - Utrecht University, 70 pp.

De Vries, B.J.M., D.P. van Vuuren and M.M. Hoogwijk, 2007. Renewable energy sources: Their global potential for the first-half of the 21st century at a global level: An integrated approach. Energy Policy 35: 2590-2610.

Dornburg, V., A. Faaij, H. Langeveld, G. van de Ven, F. Wester, H. van Keulen, K. van Diepen, J. Ros, D. van Vuuren, G.J. van den Born, M. van Oorschot, F. Smout, H. Aiking, M. Londo, H. Mozaffarian, K. Smekens, M. Meeusen, M. Banse, E. Lysen, and S. van Egmond, 2008. Biomass Assessment: Assessment of global biomass potentials and their links to food, water, biodiversity, energy demand and economy – Main Report, Study performed by Copernicus Institute – Utrecht University, MNP, LEI, WUR-PPS, ECN, IVM and the Utrecht Centre for Energy Research, within the framework of the Netherlands Research Programme on Scientific Assessment and Policy Analysis for Climate Change. Reportno: WAB 500102012, 85 pp. + Appendices.

Faaij, A., 2006. Modern biomass conversion technologies. Mitigation and Adaptation Strategies for Global Change 11: 335-367.

FAO, 2008. State of Food and Agriculture - Biofuels: prospects, risks and opportunities, Food and Agirculture Organisation of the United Nations, Rome, Italy.

Fulton, L., 2004. International Energy Agency, Biofuels for transport – an international perspective, Office of Energy Efficiency, Technology and R&D, OECD/IEA, Paris, France.

Groen, M., 1999. Energy rooted in sugar cubes; the interaction between energy savings and cogeneration in Indian sugar mills, Report of the Department of Science, Technology & Society – Utrecht University in collaboration with Winrock International, Utrecht, the Netherlands, pp. 34.

Hamelinck, C. and A. Faaij, 2006. Outlook for advanced biofuels. Energy Policy 34: 3268-3283.

Hamelinck, C.N. and A.P.C. Faaij, 2002. Future prospects for production of methanol and hydrogen from biomass. Journal of Power Sources 111: 1-22.

Hamelinck, C., A. Faaij, H. den Uil and H. Boerrigter, 2004. Production of FT transportation fuels from biomass; technical options, process analysis and optimisation and development potential. Energy, the International Journal 29: 1743-1771.

Hamelinck, C.N., G van Hooijdonk and A.P.C. Faaij, 2005. Future prospects for the production of ethanol from ligno-cellulosic biomass. Biomass & Bioenergy 28: 384-410.

Hoogwijk, M., A. Faaij, B. de Vries and W. Turkenburg, 2008. Global potential of biomass for energy from energy crops under four GHG emission scenarios Part B: the economic potential. Biomass & Bioenergy, in press.

Hoogwijk, M., A. Faaij, B. Eickhout, B.de Vries and W. Turkenburg, 2005. Potential of biomass energy out to 2100, for four IPCC SRES land-use scenarios. Biomass & Bioenergy 29: 225-257.

Hunt, S., J. Easterly, A. Faaij, C. Flavin, L. Freimuth, U. Fritsche, M. Laser, L. Lynd, J. Moreira, S. Pacca, J.L. Sawin, L. Sorkin, P. Stair, A. Szwarc and S. Trindade, 2007. Biofuels for Transport: Global Potential and Implications for Energy and Agriculture. prepared by Worldwatch Institute, for the German Ministry of Food, Agriculture and Consumer Protection (BMELV) in coordination with the German Agency for Technical Cooperation (GTZ) and the German Agency of Renewable Resources (FNR), ISBN: 1844074226, Published by EarthScan/James & James, 336 pp.

International Energy Agency, 2006. World Energy Outlook 2006, IEA Paris, France.

Kahn Ribeiro, S., S. Kobayashi, M. Beuthe, J. Gasca, D. Greene, D.S. Lee, Y. Muromachi, P.J. Newton, S. Plotkin, D. Sperling, R. Wit and P.J. Zhou, 2007: Transport and its infrastructure. In: B. Metz, O.R. Davidson, P.R. Bosch, R. Dave and L.A. Meyer (eds.), Climate Change 2007: Mitigation contribution of Working Group III to the Fourth Assessment Report of the Intergovernmental Panel on Climate Change, Cambridge University Press, Cambridge United Kingdom and New York, USA.

Larson, E.D., 2005. A Review of LCA Studies on Liquid Biofuel Systems for the Transport Sector. Energy for Sustainable Development 11: October 2005.

Lynd, L.R., W.H. van Zyl, J.E. McBride and M. Laser, 2005. Consolidated Bioprocessing of Lignocellulosic Biomass: An Update. Current Opinion in Biotechnology 16: 577-583.

Macedo, I.d.C., M.R.L.V. Leal and J.E.A.R. Da Silva, 2004. Assessment of greenhouse gas emissions in the production and use of fuel ethanol in Brazil. Secretariat of the Environment of the State of São Paulo, Brazil. Accessible at: http://www.unica.com.br/i_pages/files/pdf_ingles.pdf

OECD/FAO, 2007. Agricultural Outlook 2007-2016. Paris, France.

Smeets, E., M. Junginger, A. Faaij, A. Walter, P. Dolzan and W. Turkenburg, 2008. The sustainability of Brazilian Ethanol; an assessment of the possibilties of certified production. Biomass & Bioenergy 32: 781-813.

Tijmensen, M.J.A., A.P.C. Faaij, C.N. Hamelinck and M.R.M. van Hardeveld, 2002. Exploration of the possibilities for production of Fischer Tropsch liquids via biomass gasification. Biomass & Bioenergy 23: 129-152.

Wall-Bake, J.D., M. Junginger, A, Faaij, T. Poot and A. da Silva Walter, 2008. Explaining the experience curve: Cost reductions of Brazilian ethanol from sugarcane. Biomass & Bioenergy, in press.

Williams, R.H., E.D. Larson, R.E. Katofsky and J. Chen, J., 1995. Methanol and hydrogen from biomass for transportation, with comparisons to methanol and hydrogen from natural gas and coal. Centre for Energy and Environmental Studies, Princeton University, reportno. 292.

Chapter 8
The global impacts of US and EU biofuels policies

Wallace E. Tyner

1. Introduction

The major biofuels producers in the world are the US, EU, and Brazil. Figure 1 shows the global breakdown of biofuels production for 2006. The US and Brazil combine to produce three-fourths of global ethanol, and the EU produces three-fourths of global biodiesel. The US overtook Brazil in ethanol production, and global production now exceeds 50 billion liters. Biodiesel total production is much smaller.

In the US, Brazil, and the EU, the biofuels industries were launched with some combination of subsidies and mandates plus border protection. As production levels have grown and as oil prices have risen, all three are now switching in different degrees from reliance on subsidies to reliance on mandates. One reason is the government budget cost of subsidies, which increase as production increases. Mandates also have a cost, but it is paid by consumers at the pump assuming the biofuel is more expensive to produce than the petroleum based fuel it replaces. The consumer cost of a mandate is directly related to oil price. At low oil prices, a mandate can be expensive for consumers because high cost renewable fuel is mandated in lieu of a certain fraction of relatively lower cost petroleum. At high oil prices, the renewable fuel may even be less expensive than petroleum based fuels, so the cost can be much lower or zero.

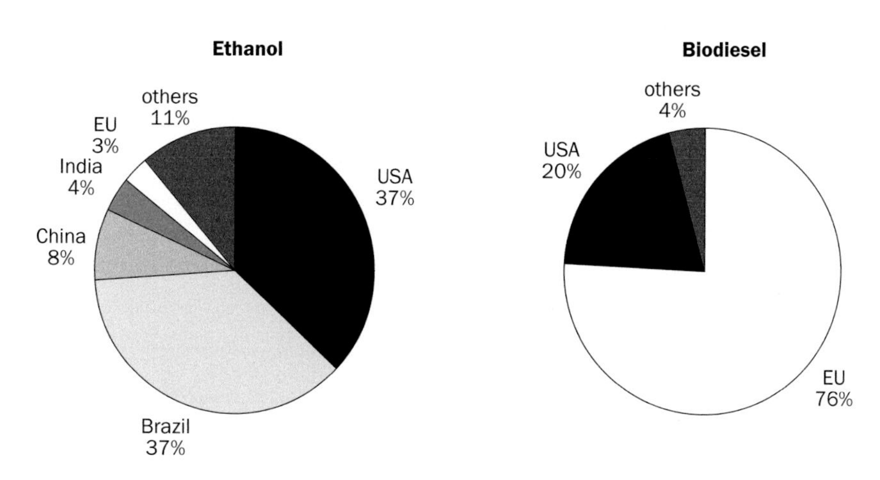

Figure 1. Global biofuels production, 2006. Data sources: Earth Policy Institute (2006), Renewable Fuels Association (2007), European Biodiesel Board (2007).

In Brazil, subsidies have been completely replaced with mandates. In the EU, subsidies are determined by each country. In essence, the EU sets a target level of renewable fuels, and each country decides how best to achieve that target. The original target was 5.75 percent renewable fuels by 2010. Most countries were well behind the pace needed to achieve that target. More recently a target of 10 percent by 2020 has been proposed. Given the recent food price and greenhouse gas controversies (more later), it appears the EU is backing away from that target. Germany has had relatively high levels of subsidics for biodiesel, but these have now ended. At present, the future directions for biofuels policies in the EU are uncertain.

In the US, ethanol has been subsidized for 30 years (Tyner, 2008). The subsidy has ranged from 10.6 to 15.9 cents per liter, and is currently 13.5 cents per liter. The subsidy on maize ethanol will be reduced to 11.9 cents per liter on 1 January 2009, but a new subsidy of 26.7 cents per liter of cellulosic ethanol will be introduced (US Congress, 2008). In addition to the subsidy, in December 2007, the US introduced biofuel mandates in the Energy Independence and Security Act (US Congress, 2007). Figure 2 portrays the timing of the US mandate, called a Renewable Fuel Standard (RFS). The Renewable Fuel Standard (RFS) as amended in the 2007 Energy Independence and Security Act calls for 36 billion gallons of renewable fuels by 2022. The RFS is divided into four categories of biofuels: conventional, advanced, cellulosic, and biodiesel. The advanced category reaches 21 billion gallons by 2022 and includes cellulosic ethanol, ethanol from sugar, ethanol from waste material, biodiesel, and other non-maize sources. In other words, the advanced category encompasses both the cellulosic and biodiesel categories. Cellulosic ethanol as a sub-set of advanced reaches 16

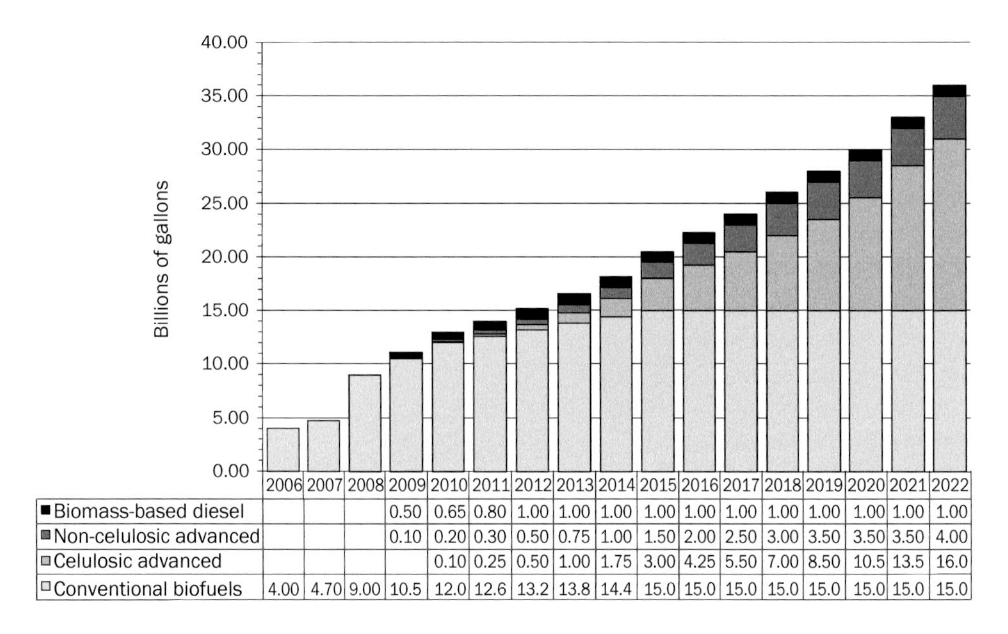

	2006	2007	2008	2009	2010	2011	2012	2013	2014	2015	2016	2017	2018	2019	2020	2021	2022
Biomass-based diesel				0.50	0.65	0.80	1.00	1.00	1.00	1.00	1.00	1.00	1.00	1.00	1.00	1.00	1.00
Non-celulosic advanced				0.10	0.20	0.30	0.50	0.75	1.00	1.50	2.00	2.50	3.00	3.50	3.50	3.50	4.00
Celulosic advanced					0.10	0.25	0.50	1.00	1.75	3.00	4.25	5.50	7.00	8.50	10.5	13.5	16.0
Conventional biofuels	4.00	4.70	9.00	10.5	12.0	12.6	13.2	13.8	14.4	15.0	15.0	15.0	15.0	15.0	15.0	15.0	15.0

Figure 2. US Renewable Fuel Standard (2007-2022). Source: Joel Valasco (pers. comm.).

billion by 2022, and biodiesel reaches 1 billion. The residual, likely to be sugarcane ethanol, amounts to 4 billion gallons by 2022. The way the standard is written, there is the total RFS requirement and the advanced requirement (with its sub-components specified separately) with the difference being presumed to be maize based ethanol. However, there is no specific RFS for maize ethanol. This residual, labeled conventional biofuels, reaches 15 billion gallons by 2015 and stays at that level. The residual is the only category that permits maize ethanol. However, it could also include any of the other categories of biofuels.

Associated with all the biofuel categories is a GHG reduction requirement. For maize based ethanol, the reduction must be at least 20 percent. For all advanced biofuels except cellulosic ethanol, the reduction required is 50 percent, and for cellulosic ethanol, it is 60 percent. Ethanol plants that were under construction or in operation as of the data of enactment of the legislation are exempt from the GHG requirement (grandfathered). The GHG requirements are to be developed and implemented by EPA. The EPA administrator has flexibility to modify to some extent the GHG percentages. S/he also has authority to reduce or waive the RFS levels.

In addition to the subsidy and RFS, the US also has a tariff on imported ethanol (Abbott *et al.*, 2008). The tariff is 2.5 percent *ad valorem* plus a specific tariff of 14.3 cents per liter of ethanol. With an ethanol CIF price of 52.9 cents per liter, the total tariff becomes 15.6 cents per liter. The rationale for the tariff was that the US ethanol subsidy applies to both domestic and imported ethanol. Congress clearly wanted to subsidize only domestically produced ethanol, so the tariff was established to offset the domestic subsidy. At the time the tariff was created, the domestic subsidy was also about 14.3 cents per liter (Tyner, 2008). However, the domestic subsidy was reduced to 13.5 and has now been reduced further to 11.9 cents per liter. Thus, today, the import tariff, as a trade barrier, goes far beyond the subsidy offset. The EU and Brazil also have import tariffs on ethanol. For Brazil, it is largely irrelevant since Brazil is one of the world's lowest cost producers of ethanol, so it is unlikely to import ethanol.

2. Ethanol economics and policy

The lowest cost ethanol source is ethanol from sugarcane. It is also the most advantageous from a net energy perspective. Brazil is the global leader in sugarcane based ethanol production, and has ample land resources to expand production. The US uses maize to produce ethanol. The cost of producing ethanol from maize varies with the price of maize. The value of the ethanol produced is a function of the price of crude oil since ethanol substitutes for gasoline. Figure 3 provides a breakeven analysis for maize ethanol at varying prices of crude oil and maize. The top line is the breakeven values with no government intervention and ethanol valued on an energy basis. The second line includes the 13.5 cent per liter subsidy. Prior to 2005, maize often ranged between $80 and $90 per mt. Without a subsidy oil would have had to be over $60 for maize ethanol to be economic. However,

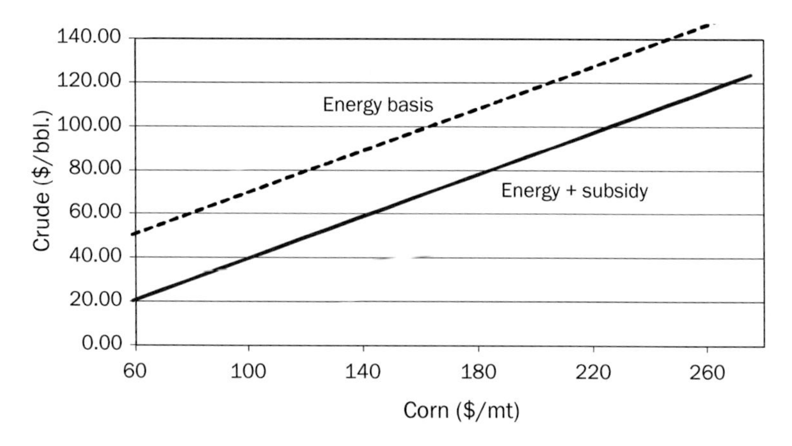

Figure 3. Breakeven ethanol prices with and without federal subsidy.

with the federal subsidy, maize ethanol was economic at around $30 crude. In addition to the federal subsidy, many US states also offered subsidies, so ethanol was attractive in the two decades prior to 2005 even though oil averaged $20/bbl. During that period It was not hugely profitable, but enough so to see the industry grow slowly over the entire period. Today with maize around $240/mt, the breakeven oil price is about $135 with no subsidy and $105 with a subsidy. The nature of a fixed subsidy is such that regardless of the maize price, the breakeven oil price difference with and without the subsidy is about $30/bbl. Or conversely, at $120 oil, the maize breakeven prices with and without subsidy are $270 and $207 per metric tonne, respectively.

2.1. Impacts of alternative US ethanol policies

This breakeven analysis is from the perspective of a representative firm. We can use a partial equilibrium economic model to examine the fixed subsidy, a variable subsidy, and the RFS over a range of oil prices (Tyner and Taheripour, 2008a,b). The model includes, maize, ethanol, gasoline, crude oil, and distillers dried grains with solubles (DDGS). The supply side of the maize market consists of identical maize producers. They produce maize using constant returns to scale Cobb-Douglas production functions and sell their product in a competitive market. Under these assumptions, we can define an aggregated Cobb-Douglas production function for the whole market. In the short-run the variable input of maize producers is a composite input which covers all inputs such as seed, fertilizers, chemicals, fuel, electricity, and so on. In short run capital and land are fixed. The demand side of the maize market consists of three users: domestic users who use maize for feed and food purposes; foreign users, and ethanol producers. We model the domestic and foreign demands with constant price elasticity functions. The foreign demand for maize is more elastic than the domestic demand. The demand of the ethanol industry for maize is a function of the demand for ethanol.

The gasoline market has two groups of producers: gasoline and ethanol producers. It is assumed that ethanol is a substitute for gasoline with no additive value. The gasoline and ethanol producers produce according to short run Cobb-Douglas production functions. The variable input of gasoline producers is crude oil and the variable input of ethanol producers is maize. Both groups of producers are price takers in product and input markets. We model the demand side with a constant price elasticity demand. The constant parameter of this function can change due to changes in income and population. We assume that the gasoline industry is well established and operates at long run equilibrium, but the ethanol industry is expanding. The new ethanol producers opt in when there are profits. There is assumed to be no physical or technical limit on ethanol production – only economic limits.

The model is calibrated to 2006 data and then solved for several scenarios. Elasticities are taken from the existing literature. Endogenous variables are gasoline supply, demand, and price: ethanol supply, demand, and price; maize price and production; maize use for ethanol, domestic use, and exports; DDGS supply and price; land used for maize; and the price of the composite input for maize. Exogenous variables include crude oil price, maize yield, ethanol conversion rate, ethanol subsidy level and policy mechanism, and gasoline demand shock (due to non-price variables such as population and income). The model is driven and solved by market clearing conditions that maize supply equal the sum of maize demands and that ethanol production expands to the point of zero profit. The model is simulated over a range of oil prices between $40 and $140.

Figure 4 provides the results from this model simulation for maize price and Figure 5 for ethanol production. In each figure, the far left bar is the 13.5 cent fixed subsidy, the second is no subsidy, the third a subsidy that varies with the price of crude oil, the fourth the RFS alone, and the fifth the RFS in combination with the fixed subsidy (current policy). The variable subsidy is in effect only for crude oil prices below $70. The first thing to note from Figure 4 is that, just as was evident from the perspective of the firm, there is now a tight linkage between crude oil price and maize price. The basic mechanism is that gasoline price is driven by crude price. Ethanol is a close substitute for gasoline, so a higher gasoline price means larger ethanol demand. That demand stimulates investment in ethanol plants. More ethanol plants means greater demand for maize, and that increased demand means higher maize price. This is a huge change, as historically, there was very little correlation between energy and agricultural prices.

The $40 oil price represents the approximate price in 2004. The model accurately 'predicts' the ethanol production and maize price corresponding to $40 oil. That is, the 2004 model results are very close to the actual 2004 values. The ethanol production under no subsidy also accurately shows ethanol production beginning only when oil reaches $60 and then at a very low level. Of course, the RFS case has the ethanol production level at 56.7 bil. l., which is the level of the RFS in 2015, and the level modeled in this analysis. The numbers above the RFS bar in Figure 5 represent the implicit subsidy on ethanol ($/gal. ethanol) due to the

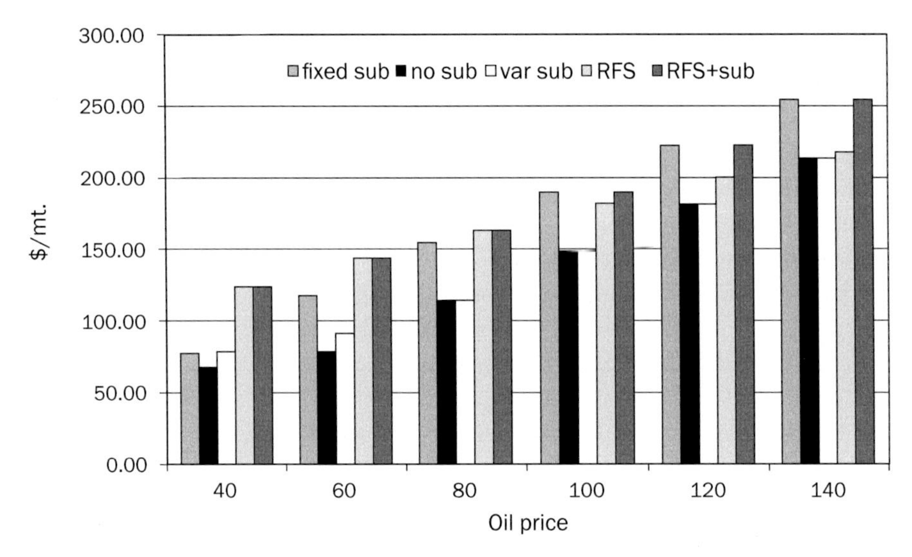

Figure 4. Maize price under alternative policies and oil prices.

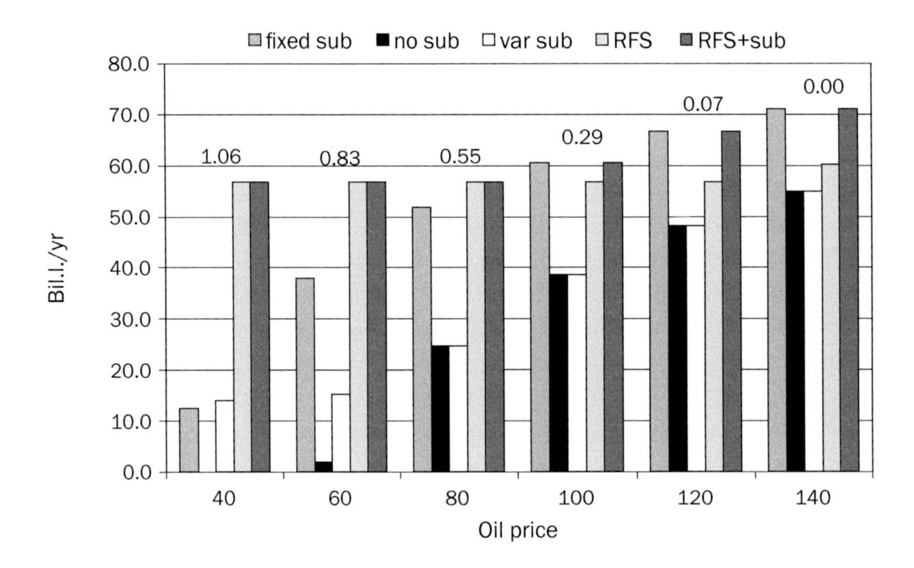

Figure 5. Ethanol production under alternative policies and oil prices.

RFS. It is also an implicit tax on consumers. The model follows the RFS rule, and 'requires' that the stipulated level of ethanol be produced. To the extent that the cost of ethanol is higher than the cost of gasoline, this higher cost gets passed on to consumers in the form of an implicit tax on consumers. Thus, a RFS functions very differently from a subsidy. The subsidy is on the government budget, whereas the mandate cost is paid by consumers directly at the pump. When oil is very inexpensive, the ethanol costs considerably more than petroleum. So the requirement to blend ethanol means consumers pay more at the pump than they would without the mandate. For $40 oil, the implicit subsidy/tax is $1.06/gal. or 28 cents per liter. The subsidy/tax falls to zero at $140 oil. At $140 oil, the mandate is no longer binding, and the amount of ethanol demanded is market driven – not determined by the mandate. Thus the RFS is a form of variable subsidy for the ethanol producer and variable tax for the consumer depending on the price of crude oil. Ethanol production stays at the RFS level of 56.7 bil. l. until oil reaches $120. At that oil price and beyond the market demands more than 56.7 bil. l., and the RFS becomes non-binding.

The final bar is the current policy of RFS plus subsidy. Note that at low oil prices, the RFS production level is higher than that induced by the subsidy, and at high oil prices, the subsidy induces higher production than the RFS. If the RFS represents the intent of Congress with respect to level of ethanol production, the subsidy takes production well beyond that level at high oil prices.

Another important question that can be addressed using these model results is what proportion of the maize price increase is due to the oil price increase, and what proportion to the subsidy. If we start at the no subsidy case with $40 oil, we have a maize price of $67, which increases to $181 when oil triples to $120. If we add on the subsidy at $120 oil, the maize price goes up to $222. The total maize price increase is $155, of which $41 is due to the subsidy, and $113 to the oil price increase. So roughly ¾ of the maize price increase has been due to higher oil prices, and ¼ to the US subsidy on maize ethanol. Even if the subsidy went away, maize prices would not return to their historic levels because of the new link between energy and agriculture. And if oil price went down, we would expect to see the maize price fall as well. As the oil price fell, gasoline would fall as would the price of ethanol. With lower ethanol prices, some plants could not produce profitably, so maize demand would fall and also the maize price.

Figure 6 displays the annual costs of the various policy options. Recall that the method of paying the costs is very different between the government subsidy and the RFS. The RFS is paid by the consumer at the pump, and the fixed and variable subsidies are paid through the government budget. The variable subsidy has no cost for oil above $70 by design, and its cost at low oil prices is quite low. The cost of the fixed subsidy increases almost linearly with oil price. The higher the oil price, the higher the government subsidy cost. The RFS is exactly opposite. It has a high cost when oil price is low, and a very low or zero cost at high oil prices.

The US tariff on imported ethanol introduces a potentially greater distortion than does the subsidy or mandate. Since high oil prices directly lead to higher maize prices, maize ethanol becomes much more expensive. Sugarcane-based ethanol is less expensive to produce than maize ethanol at any oil price, but the gap widens at higher oil prices. So removal

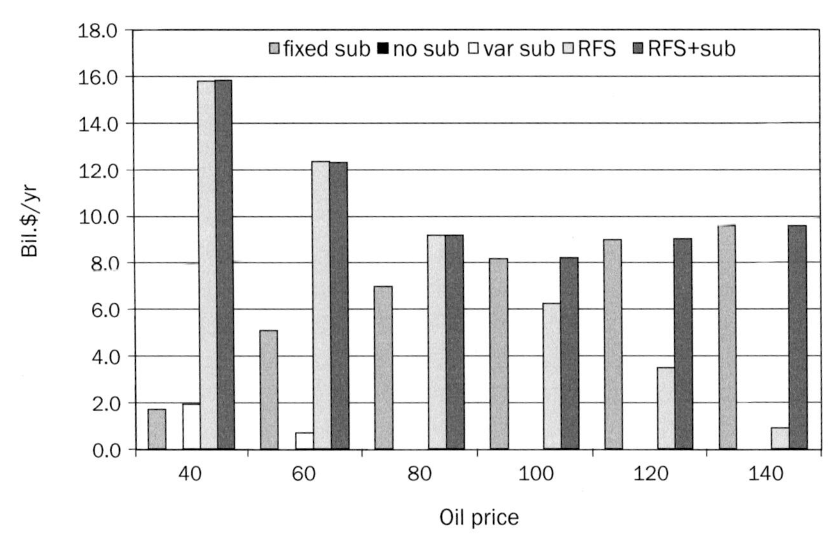

Figure 6. Costs of the policy alternatives.

of the tariff on imported ethanol would lead to the biofuel coming from the lowest cost source–sugarcane–which would reduce some pressure on maize prices and provide the United States with lower cost ethanol. Brazil has the potential to expand ethanol production substantially without increasing world sugar prices substantially, so imports down the road could be quite high.

However, the question is more complicated because it depends on the extent to which imported ethanol adds to total consumption and the extent to which it displaces maize ethanol. For the portion that displaced maize ethanol, each billion gallons of imports would displace about 358 million bushels of maize used for ethanol (Tyner and Taheripour, 2007). So you would get price impacts as the ethanol industry demanded less maize. The problem is figuring out how much would go to increase total consumption and how much to displace maize ethanol. In the United States, the limit of how much ethanol can be blended is called the blending wall (Tyner *et al.*, 2008). The blending wall is the maximum amount of ethanol that can be blended at the regulatory maximum of 10%. Currently, we consume about 140 billion gallons of gasoline (Energy Information Administration, 2008), so the max level for the blending wall would be 14 billion gallons of ethanol. However, for logistical reasons, the practical level is likely to be much lower, perhaps around 12 billion gallons. See Tyner *et al.* (2008) for a more complete analysis of this issue.

We already have in place or under construction 13 billion gallons of ethanol capacity. At present E85 is tiny, and it would take quite a while to build that market. There are only about 1,700 E85 pumps in the nation and few flex-fuel vehicles that are required to consume the fuel. It would require a massive investment to make E85 pumps readily available for all consumers, and a huge switch to flex-fuel vehicle manufacture and sale to grow this market. Without strong government intervention, it will not happen.

What options exist? The most popular among the ethanol industry is switching to E15 or E20 instead of E10. The major problem is that automobile manufacturers believe the existing fleet is not suitable for anything over E10. Switching to a higher blend would void warranties on the existing fleet and potentially pose problems for older vehicles not under warranty. In the US, the automobile fleet turns over in about 14 years, so it is a long term process. We could not add yet another pump for E15 or E20. The costs would be huge. So the blending wall in the near term is an effective barrier to growth of the ethanol industry. If a switch is made to an E15 or E20 limit for standard cars, some agreement would have to be reached on who pays for any vehicle repair or performance issues.

On the technical side, two options could emerge. One would be using cellulose through a thermochemical conversion process to produce gasoline or diesel fuel directly. Today this process is quite expensive, but the cost might be reduced over the next few years. A second option is to convert cellulose to butanol instead of ethanol, which is much more similar to gasoline. Without such a breakthrough, the EPA administrator likely will be forced to cap the RFS far below the planned levels.

Until we hit the blending wall, most of the imports likely would increase total consumption and not displace maize ethanol. However, we will probably reach the blending wall in 2009/10, at which point imports would likely displace domestic maize ethanol and thereby lower maize price.

3. Impacts of US and EU policies on the rest of the world

Our analysis of global impacts is done using the Global Trade Analysis Project (GTAP) model and data base. This work is based on Hertel *et al.* (2008). We begin with an analysis of the origins of the recent bio-fuel boom, using the historical period from 2001-2006 for purposes of model calibration and validation. This was a period of rapidly rising oil prices, increased subsidies in the EU, and, in the US, there was a ban on the major competitor to ethanol for gasoline additives (MTBE) (Tyner, 2008). Our analysis of this historical period permits us to evaluate the relative contribution of each of these factors to the global biofuel boom. We also use this historical simulation to establish a 2006 benchmark biofuel economy from which we conduct our analysis of future mandates.

We then can do a forward-looking analysis of EU and US biofuel programs. The US Energy Policy and Security Act of 2007 calls for 15 billion gallons of ethanol use by 2015, most of which is expected to come from maize. In the EU, the target is 5.75% of renewable fuel use in 2010 and 10% by 2020. However, there are significant doubts as to whether these goals are attainable. For this analysis, we adopt the conservative mandate of 6.25% by 2015 in the EU.

The starting point for our prospective simulations is the updated, 2006 fuel economy which results from the foregoing historical analysis. Thus, we analyze the impact of a continued intensification of the use of biofuels in the economy treating the mandates as exogenous shocks.[12] Ethanol exports from Brazil to the US grow in this simulation as well.

Table 1 reports the percentage changes in output for biofuels and the land-using sectors in the USA, EU and Brazil. The first column in each block corresponds to the combined impact of EU and US policies on a given sector's output (USEU-2015). The second column in each block reports the component of this attributable to the US policies (US-2015), and the third reports the component of the total due to the EU policies (EU-2015) using the decomposition technique of Harrison *et al.* (2000). This decomposition method is a more sophisticated approach to the idea of first simulating the global impacts of a US program, then simulating the impact of an EU biofuels program, and finally, simulating the impact of the two combined. The problem with that (rather intuitive) approach is that the impacts of the individual programs will not sum to the total, due to interactions. By adopting this numerical integration approach to decomposition, the combined impacts of the two programs are fully attributed to each one individually.

In the case of the US impacts (columns labeled Output in US), most of the impacts on the land-using sectors are due to US policies. Coarse grains output rises by more than 16%, while output of other crops and livestock falls when only US policies are considered. However, oilseeds are a major exception. Here, the production impact is reversed when EU mandates are introduced. In order to meet the 6.25% renewable fuel share target, the EU requires a massive amount of oilseeds. Even though production in the EU rises by 52%, additional imports of oilseeds and vegetable oils are required, and this serves to stimulate production worldwide, including in the US. Thus, while US oilseeds output falls by 5.6% in the presence of US-only programs, due to the dominance of ethanol in the US biofuel mix, when the EU policies are added to the mix, US oilseed production actually rises.

In the case of the EU production impacts (Output in EU: the second group of columns in Table 1), the impact of US policies is quite modest, with the main interaction again through the oilseeds market. However, when it comes to third markets – in particular Brazil (Output in Brazil), the US and EU both have important impacts. US policies drive sugarcane production, through the ethanol sector, while the EU policies drive oilseeds production in Brazil. Other crops, livestock, and forestry give up land to these sectors.

[12] Technically, we endogenize the subsidy on biofuel use and exogenize the renewable fuel share, then shock the latter. For simplicity, all components of the renewable fuels bundle are assumed to grow in the same proportion.

Table 1. Change in output due to EU and US biofuel mandates: 2006-2015 (%).

Sector	Output in US			Output in EU			Output in Brazil		
	USEU-2015	**US-2015**	**EU-2015**	**USEU-2015**	**US-2015**	**EU-2015**	**USEU-2015**	**US-2015**	**EU-2015**
Ethanol	177.5	177.4	0.1	430.9	1.3	429.7	18.1	17.9	0.2
Biodiesel	176.9	176.8	0.1	428.8	1.2	427.6	-	-	-
Coarse grains	16.6	16.4	0.2	2.5	0.8	1.7	-0.3	1.1	-1.4
Oilseeds	6.8	-5.6	12.4	51.9	1.2	50.7	21.1	0.6	20.5
Sugarcane	-1.8	-1.9	0.1	-3.7	0.0	-3.7	8.4	9.3	-0.9
Other grains	-7.6	-8.7	1.2	-12.2	0.1	-12.3	-8.7	-2.0	-6.8
Other agri	-1.6	-1.7	0.2	-4.5	0.0	-4.5	-3.8	-1.5	-2.4
Livestock	-1.2	-1.2	0.0	-1.7	0.1	-1.8	-1.4	-0.6	-0.7
Forestry	-1.2	-1.4	0.1	-5.4	-0.3	-5.1	-2.7	-1.0	-1.8

Note: Ethanol in the US and EU is from grains, and it is sugarcane-based in Brazil.

Table 2 reports changes in crop harvested area as a result of the biofuel mandates in the US and EU for all regions in the model. The simulation includes only the biofuels shock, and does not include population growth, income growth, trend yield increases, or anyother 'baseline' factors. It is designed just to isolate the biofuels impacts. Coarse grains acreage in the US is up by about 10%, while sugar, other grains, and other crops are all down. The productivity-weighted rise in coarse grains acreage is 10% (Table 3). This increase in maize acreage in the US comes from contribution of land from other land-using sectors such as other grains (Table 3) as well as pasture land and commercial forest land – to which we will turn momentarily.

From Table 2, we see that US oilseeds acreage is up slightly due to the influence of EU policies on the global oilseeds market. However, this marginal increase is dwarfed by the increased acreage devoted to oilseeds in other regions, where the percentage increases range from 11 to 16% in Latin America, and 14% in Southeast Asia and Africa, to 40% in the EU. If the EU really intends to implement its 2015 renewable fuels target, there will surely be a global boom in oilseeds. Coarse grains acreage in most other regions is also up, but by much smaller percentages. Clearly the US-led ethanol boom is not as significant a factor as the EU oilseeds boom. Sugarcane area rises in Brazil, but declines elsewhere, and other grains and crops are somewhat of a mixed bag, with acreage rising in some regions to make up for diminished production in the US and EU and declines elsewhere.

From an environmental point of view, the big issue is not which crops are grown, but how much cropland is demanded overall, and how much (and where) grazing and forestlands are converted to cropland. These results are very sensitive to the productivity of land in the pasture and forest categories compared to cropland. We recognize that more work needs to be done on certain land categories such as idled land and cropland pasture in the US and the savannah in Brazil. Therefore the numerical results reported here must be taken as only illustrative of the results that will be available once the land data base is improved. Table 3 reports the percentage changes in different land cover areas as a result of the EU and US mandates. Furthermore, as with the output changes in Table 1, we decompose this total into the portion due to each region's biofuels programs. From the first group of columns, we see that crop cover is up in nearly all regions. Here we also see quite a bit of interaction between the two sets of programs. For example, in the US, about one-third of the rise in crop cover is due to the EU programs. In the EU, the US programs account for a small fraction of the rise in crop cover. In other regions, the EU programs play the largest role in increasing crop cover. For example, in Brazil, the EU programs account for nearly 11% of the 14.2% rise in crop cover.

Where does this crop land come from? In our framework it is restricted to come from pastureland and commercial forest lands, since we do not take into account idle lands, nor do we consider the possibility of accessing currently inaccessible forests. The largest percentage reductions tend to be in pasturelands (Table 3, final set of columns). For example, in Brazil,

Table 2. Change in crop harvested area by region, due to EU and US biofuel mandates: 2006-2015 (%).

Region	**Crops**				
	Coarse grains	**Oilseeds**	**Sugarcane**	**Other grains**	**Other agri**
USA	9.8	1.6	-5.7	-10	-2.7
Canada	3.5	16.9	-3.2	-2.6	-1.6
EU-27	-2.3	40	-7.4	-15.1	-6.1
Brazil	-3.2	16	3.8	-10.9	-5.1
Japan	10.7	7.6	-0.7	0.8	-0.1
China-Hong Kong	1.2	8.2	-0.6	-0.5	-0.5
India	-0.7	0.9	-0.7	0.5	-0.2
Latin American energy exporters	1.8	11.3	-2.3	-0.2	-0.8
Rest of Latin America & Caribbean	1.7	11.5	-1.6	-0.6	-0.3
EE & FSU energy exporters	0.5	18.1	-0.6	0.4	-0.5
Rest of Europe	2.3	10.5	0	1.8	0.4
Middle Eastern North Africa energy exporters	4	8.6	-0.9	2.5	-0.4
Sub Saharan energy exporters	-0.8	13.7	0	2.3	1.2
Rest of North Africa & SSA	1.5	14.2	-0.4	1.1	1.1
South Asian energy exporters	-0.5	3.7	-0.9	-0.6	-0.1
Rest of high income Asia	3.7	6.1	-0.1	-0.2	0
Rest of Southeast & South Asia	-0.2	2.9	-0.8	0	-0.1
Oceania countries	3.9	17.2	-0.6	-1.3	0.3

Note: These results are solely illustrative of the kinds of numerical results that are produced by the analysis. They are not definitive results.

we estimate that pasturelands could decline by nearly 11% as a result of this global push for biofuels, of which 8% decline is from EU mandates alone. The largest percentage declines in commercial forestry cover are in the EU and Canada, followed by Africa. In most other regions, the percentage decline in forest cover is much smaller.

Our prospective analysis of the impacts of the biofuels boom on commodity markets focused on the 2006-2015 time period, during which existing investments and new mandates in the US and EU are expected to substantially increase the share of agricultural products (e.g. maize in the US, oilseeds in the EU, and sugar in Brazil) utilized by the biofuels sector. In

Table 3. Decomposition of change land cover by EU and US biofuel mandates (with Sensitivity Analysis): 2006-2015 (% change).

	Crop cover				
	USEU 2015	US 2015	EU 2015	Confidence interval (95%)	
				Lower	Upper
US	7	4.7	2.3	3.5	10.8
Canada	11.3	2.9	8.4	4.7	18.0
EU-27	14.3	0.9	13.4	8.0	20.7
Brazil	14.2	3.5	10.7	7.0	21.5
Japan	1.3	0.5	0.8	-0.1	2.7
China-Hong Kong	1.9	0.5	1.4	-0.5	4.3
India	1	0.1	0.9	-0.6	2.7
Latin American EEx.	6.2	2.1	4.1	1.6	10.9
Rest of Latin Am.	5.5	1.5	4.1	1.3	9.9
EE & FSU EEx.	4.6	0.9	3.7	0.1	9.1
Rest of Europe	6.8	1.3	5.5	2.1	11.5
Middle Eastern N Africa EEx.	1.7	0.4	1.2	0.2	3.2
Sub Saharan EEx.	6.9	1.6	5.3	1.7	12.1
Rest of North Africa & SSA	9.9	2.1	7.8	3.3	16.6
South Asian EEx.	-0.2	0	-0.2	-0.9	0.5
Rest of high income Asia	0.1	0	0	-0.1	0.2
Rest of Southeast & South Asia	1.2	0.2	1	-0.3	2.7
Oceania countries	6.6	1.5	5.1	1.6	11.7

the US, this share could more than double from 2006 levels, while the share of oilseeds going to biodiesel in the EU could triple. In analyzing the biofuel policies in these regions, we decompose the contribution of each set of regional policies to the global changes in output and land use. The most dramatic interaction between the two sets of policies is for oilseed production in the US, where the sign of the output change is reversed in the presence of EU mandates (rising rather than falling). The other area where they have important interactions is in the aggregate demand for crop land. About one-third of the growth in US crop cover is attributed to the EU mandates. When it comes to the assessing the impacts of these mandates on third economies, the combined policies have a much greater impact than just the US or just the EU policies alone, with crop cover rising sharply in Latin America, Africa

Forest cover					Pasture cover				
USEU 2015	US 2015	EU 2015	Confidence interval (95%)		USEU 2015	US 2015	EU 2015	Confidence interval (95%)	
			Lower	Upper				Lower	Upper
-1.7	-1.3	-0.5	-2.6	-0.9	-4.9	-3.2	-1.7	-7.3	-2.6
-6	-1.6	-4.4	-9.2	-2.8	-4.4	-1.1	-3.4	-6.9	-2.1
-7.3	-0.5	-6.8	-10.4	-4.3	-5.6	-0.4	-5.3	-7.8	-3.5
-1.7	-0.5	-1.2	-2.5	-0.9	-10.9	-2.7	-8.3	-15.8	-6.1
-0.8	-0.3	-0.5	-1.8	0.2	-0.4	-0.2	-0.3	-0.8	-0.1
0.1	0	0.2	-0.2	0.5	-2	-0.4	-1.6	-4.1	0.1
0	0	0	-0.4	0.4	-1	-0.1	-0.9	-2.4	0.3
-2	-0.8	-1.2	-3.3	-0.6	-4	-1.3	-2.7	-6.8	-1.2
-0.3	-0.3	0	-1.5	0.9	-5	-1.1	-3.9	-8.3	-1.7
-0.8	-0.2	-0.5	-3.6	2.0	-3.6	-0.6	-3	-6.0	-1.2
-0.7	-0.3	-0.4	-2.0	0.7	-5.7	-0.9	-4.8	-9.2	-2.3
-0.9	-0.2	-0.6	-1.7	0.0	-0.8	-0.2	-0.6	-1.4	-0.2
-3.4	-0.8	-2.6	-6.3	-0.5	-3.2	-0.7	-2.5	-5.1	-1.2
-3.4	-0.8	-2.6	-5.8	-1.1	-5.8	-1.1	-4.6	-9.2	-2.4
0.5	0.1	0.4	-0.2	1.2	-0.3	-0.1	-0.2	-0.5	0.0
0.1	0	0.1	0.0	0.2	-0.1	0	-0.1	-0.3	0.0
0	0	0	-0.3	0.2	-1.1	-0.2	-0.9	-2.5	0.2
-2.4	-0.6	-1.8	-4.0	-0.8	-3.9	-0.8	-3.1	-6.8	-1.0

and Oceania as a result of the biofuel mandates. These increases in crop cover come at the expense of pasturelands (first and foremost) as well as commercial forests. It is these land use changes that have attracted great attention in the literature (e.g. Searchinger *et al.*, 2008) and a logical next step would be to combine this global analysis of land use with estimates of the associated greenhouse gas emissions.

4. Conclusions

This paper examines US ethanol policy options using a partial equilibrium model and US and EU options using a global general equilibrium model. The partial equilibrium results clearly illustrate the new linkage between energy and agricultural markets. Prices of agricultural commodities in the future will be driven not only by demand and supply relationships for the agricultural commodities themselves, but also by the price of crude oil. Ethanol from maize and sugarcane can be produced economically at high crude oil prices. The US policy interventions have enabled the ethanol industry to exist and grow over the past 30 years. Today the government interventions continue to be important, but the new added driver is high oil prices.

When one examines the US and EU policies together, one sees clearly that the impacts are felt around the world. Trade and production patterns are affected in every region. The results presented here are very preliminary, but they serve to illustrate how the analysis can be used to estimate global production, trade, and land use impacts of US and EU policies.

Acknowledgements

The author acknowledges the collaboration of Dileep Birur, Tom Hertel, and Farzad Taheripour.

References

Abbott, P., C. Hurt and W. Tyner, 2008. What's Driving Food Prices? Farm Foundation Issue Report, July 2008. Available at: www.farmfoundation.org.

Earth Policy Institute, 2006. World Ethanol Production. Available at: www.earthpolicy.org/Updates/2005/Update49_data.htm.

Energy Information Administration, 2008. U.S. Department of Energy. Available at: www.eia.doe.gov.

European Biodiesel Board, 2007. The EU Biodiesel Industry. Available at: www.ebb-eu.org/stats.php.

Harrison, W.J., J.M. Horridge and K.R. Pearson, 2000 Decomposing Simulation Results with Respect to Exogenous Shocks. Computational Economics 15: 227-249.

Hertel, T.W., W.E. Tyner and D.K. Birur, 2008 Biofuels for all? Understanding the Global Impacts of Multinational Mandates GTAP Working Paper No. 51, 2008.

Renewable Fuels Association, 2007. World Fuel Ethanol Production. Available at: www.ethanolrfa.org/industry/statistics/#A.

Searchinger, T., R. Heimlich, R.A. Houghton, F. Dong, A. Elobeid, J. Fabiosa, S. Tokgoz, D. Hayes and T.-H. Yu, 2008. Use of U.S. Croplands for Biofuels Increases Greenhouse Gases Through Emissions from Land-Use Change. Science 317: 1238-1240.

Tyner, W.E., 2008. The US Ethanol and Biofuels Boom: Its Origins, Current Status, and Future Prospects. BioScience 58: 646-53.

Tyner, W.E. and F. Taheripour, 2007. Future Biofuels Policy Alternatives. paper presented at the Farm Foundation/USDA conference on, April 12-13, 2007, St. Louis, Missouri. In: Biofuels, Food, and Feed Tradeoffs, pp. 10-18. Available at: http://www.farmfoundation.org/projects/documents/Tynerpolicyalternativesrevised4-20-07.pdf.

Tyner, W. and F. Taheripour, 2008a. Policy Options for Integrated Energy and Agricultural Markets. Review of Agricultural Economics 30, in press.

Tyner, W. and F. Taheripour, 2008b. Policy Analysis for Integrated Energy and Agricultural Markets in a Partial Equilibrium Framework. Paper Presented at the Transition to a Bio-Economy: Integration of Agricultural and Energy Systems conference on February 12-13, 2008. Available at: www.agecon.purdue.edu/papers.

Tyner, W.E., F. Dooley, C. Hurt and J. Quear, 2008. Ethanol Pricing Issues for 2008. Industrial Fuels and Power, February 2008: 50-57.

U.S. Congress, 2007. Energy Independence and Security Act of 2007. H.R. 6, 110 Congress, 1st session. Available at: http://www.whitehouse.gov/news/releases/2007/12/20071219-1.html.

U.S. Congress, 2008. 2008 Farm Bill. H. R. 6124 (P.L. 110-246).

Chapter 9
Impacts of sugarcane bioethanol towards the Millennium Development Goals

Annie Dufey

1. Introduction

At the Millennium Summit in September 2000 the largest gathering of world leaders in history adopted the United Nations Millennium Declaration. They committed to a new global partnership to reduce extreme poverty by 2015 in line with a series of targets that have become known as the Millennium Development Goals (MDGs). The MDGs are crafted around eight themes to promote sustainable development addressing extreme poverty in its different dimensions including hunger, health, education, the promotion of gender quality and environmental sustainability (see Box 1).

At the same time, during the last five years or so, the world has witnessed the global emergence of a new sector – the biofuels sector. Biofuels potential for achieving simultaneously economic, poverty reduction and environmental goals have combined and placed biofuels at the top of today's most pressing policy agendas.

This chapter argues that sugarcane bioethanol can be supportive of sustainable development and poverty reduction, thus contributing to the achievement of the MDGs. In some contexts there might be synergies between the pursue of different goals but there may be

Box 1. The Millennium Development Goals.

The eight Millennium Development Goals were agreed at the United Nations Millennium Summit in September 2000. The eight Millennium Development Goals are:

- Eradicate extreme poverty and hunger
- Achieve universal primary education
- Promote gender equality and empower women
- Reduce child mortality
- Improve maternal health
- Combat HIV and AIDS, malaria and other diseases
- Ensure environmental sustainability
- Develop a global partnership for development

Source: http://www.un.org/millenniumgoals/

also risks and serious trade-offs over food security, small farmers inclusion, environment and the economy.

Much of the available evidence comes from Brazil, which has the main longstanding experience with the launching of the PROALCOOL Programme in 1975 to replace imported gasoline with bioethanol produced from locally grown sugarcane. Today Brazil is the second bioethanol producer after the United States and the main exporter. In addition, there have been other smaller initiatives with different rate of success. These include African and East Asian countries such as Zimbabwe, Malawi, Kenya, Pakistan and India that have promoted bioethanol from sugarcane molasses, some of them since the early eighties. More widely, at present, many countries around the world, in their search for development and poverty reduction opportunities are trying to replicate the Brazilian experience with sugarcane bioethanol. Their vast majority are developing countries in tropical and semitropical areas in the Caribbean, Africa, Latin America and East Asia in which sugarcane is traditionally grown.

The chapter is organized as follows. After this brief introduction, Section 2 argues that sugarcane bioethanol may offer some genuine opportunities for sustainable development and poverty reduction and identify the key potential benefits. Section 3 points out that benefits are not straightforward and identifies several challenges and trade-offs that need to be confronted in order to realize their full potential for achieving sustainable development and poverty reduction. Finally, section 4 concludes and provides some recommendations.

2. Opportunities for sugarcane bioethanol in achieving sustainable development and the Millennium Development Goals

Sugarcane bioethanol can contribute to sustainable development and poverty reduction through a varied range of environmental, social and economic advantages over fossil fuels. These include: (a) enhanced energy security both at national and local level; (b) improved social well-being through better energy services especially among the poorest; (c) improved trade balance by reducing oil imports; (d) rural development and better livelihoods; (e) product diversification leaving countries better-off to deal with market fluctuations; (f) creation of new exports opportunities; (g) potential to help tackling climate change through reduced emissions of greenhouse gases (h) reduced emissions of other air contaminants; and (i) opportunities for investment attraction through the carbon finance markets. This section briefly addresses each of these aspects.

2.1. Enhanced energy security

Enhanced energy security has become a universal geopolitical policy concern and it was a key policy driver behind the first attempts to introduce sugarcane bioethanol at a massive scale in the mid-1970s in Brazil (Dufey *et al.*, 2007b). Current increasing energy costs and uncertainty regarding future energy supply are giving many governments incentive to encourage the production of petroleum substitutes from agricultural commodities. Indeed, the volatility of world oil prices, uneven global distribution of oil supplies, uncompetitive structures governing the oil supply and heavy dependence on imported fuels are all factors that leave many countries vulnerable to disruption of supply, imposing serious energy security risks which can result in physical hardships and economic burden (Dufey, 2006). For instance, crude oil imports to African, Caribbean and Pacific countries were expected to increase to 72 percent of their requirements in 2005 (Coelho, 2005).

Energy diversification makes countries less vulnerable to oil price shocks, compromising macro-stability affecting variables such as the exchange rate, inflation and debt levels (Cloin, 2007). Sugarcane bioethanol is a rational choice in countries where sugarcane can be produced at reasonable cost without adverse social and environmental impacts (Dufey *et al.*, 2007b). For remote places, locally produced sugarcane bioethanol can offer a highly competitive alternative to other fuels. This might be the case of several sugarcane producing countries in Pacific island nations and land-locked countries in Africa where the high costs of fossil fuel transportation and the related logistics make them prohibitive.

2.2. Benefits at the household level - improved social well-being

A large part of the poor, mostly in rural areas, do not have access to affordable energy services which affects their chances of benefiting from economic development and improved living standards. In this context the use of bioethanol and other renewable sources can directly or indirectly lead to several MDGs including gender equality, reduction of child mortality, poverty reduction, improvement of maternal health and environmental sustainability. Firstly, they can reduce the time spent by women and children on basic survival activities (gathering firewood, fetching water, cooking, etc.). Women in least developed countries may spend more than one third of their productive life collecting and transporting wood. Additional help needed from children often prevents them from attending school (FAO, 2007). Secondly, the use of bioethanol (and other liquid biofuels) for household cooking and heating could help to reduce respiratory disease and death associated with burning of other traditional forms of fuels usually used in the poorest countries (e.g charcoal, fuelwood and paraffin solid biomass fuels indoors), to which women and children are especially vulnerable (UN-Energy, 2007; Woods and Read, 2005). In some African countries charcoal and woodfuel account for over 95 percent of household fuel (Johnson and Rosillo-Calle, 2007). As Box 2 suggests, experiences promoting the use of sugarcane bioethanol in stoves at the household level are expected to report important socio-economic and environmental benefits. Finally, the use of biofuels can improve access to pumped drinking water, which can reduce hunger by allowing for cooked food (95% of food needs cooking) (Gonsalves, 2006a). However, adaptation of bioethanol for domestic uses would of course require a cultural shift away from the traditional hearth, plus attention to safety in fuel storage, as liquid biofuels are highly flammable (Dufey *et al.*, 2007b). Overall, electricity through transmission lines to

Box 2. Bioethanol stoves to condominium residents in Addis Ababa in Ethiopia

In Ethiopia the Municipality of Addis Ababa EPA (Environmental Protection Authority) and a Sub-City district are working closely with Gaia Association, Dometic AB, Makobu Enterprises, and Finchaa Sugar Factory to develop a project whereby initially 2000 CleanCook (CC) stoves will be installed in newly built condominium apartments. Wood and charcoal stoves are not permitted in these condominium buildings.
The CC stove is financed within the condominium unit price. Financing is provided by the condominium association with the assistance of the Municipal EPA, the Sub-City Administration and a financing entity. The finance rate is regulated by the government and is kept low. The bioethanol used in the project is produced at one of three state-owned sugar factories at a contractual price by Makobu Enterprises and delivered to the condominium. The fuel storage and distribution infrastructure will be financed by the condominium association. The Ethiopian EPA will work with one Sub-City Administration to package the stove financing into the condominium financing through the national bank. As a result, 2000 CC stoves will be financed in 2008 and approximately 360,000 liters of domestically produced bioethanol will supplant kerosene, charcoal and firewood use. The other nine Sub-City administrations could replicate the model.
Since the CC stove is clean burning, its introduction will improve indoor air quality and, consequently, household health. Another advantage of this model lies in the potential for Clean Development Mechanism (CDM) financing. It is important to note the government has had a central role for the development of a domestic bioethanol industry in Ethiopia, as well as for building a local market for bioethanol as a household cooking fuel. Indeed, after considering allocating bioethanol for fuel blending in the transport sector in 2006, the Government got convinced that the most significant socioeconomic and environmental benefits would stem from prioritizing the use in the domestic household sector.

Source: adapted from Lambe (2008).

many rural areas is unlikely to happen in the near future, so access to modern decentralized small-scale energy technologies, particularly renewables are an important element for effective poverty alleviation policies (Gonsalves, 2006a). In this context, bioethanol can be directed towards high value added uses such as lighting or motors, which can lead to income generating activities.

But the effectiveness of using sugarcane bioethanol for these uses would need to be assessed against those of other energy crops or renewable sources such as small hydropower.

2.3. Improved trade balance

Heavy reliance on foreign energy sources means countries have to spend a large proportion of their foreign currency reserves on oil imports. Oil import dependency is especially acute in Sub-Saharan and East Asian countries, where 98 percent and 85 percent of their oil needs are met by imports, respectively (ESMAP, 2005a). Changes in oil prices have devastating effects in these countries. For instance, the 2005 oil price surge reduced Gross Domestic Product growth of net oil importing countries from 6.4 percent to 3.7 percent, and, as a consequence, the number of people in poverty rose by as much as 4-6 percent, with nearly 20 countries experiencing increases of more than 2 percent (ESMAP, 2006).

Domestically produced bioethanol offers oil importing countries an opportunity to improve their trade balance. In Brazil, for instance, the replacement of imported gasoline by sugarcane bioethanol saved the country some US$ 61 billion in avoided oil imports during the last eight years – equating the total amount of the Brazilian external public debt (FAO, 2007). In Colombia, the implementation of the bioethanol programme would result in foreign exchange savings of US$ 150 million a year (Echeverri-Campuzano, 2000).

2.4. Rural development and creation of sustainable livelihoods

Biofuels provide new economic opportunities and employment in the agricultural sector, key aspects for poverty reduction. They generate a new demand for agricultural products that goes beyond traditional food, feed and fibre uses, expanding domestic markets for agricultural produce and paving the way for more value-added produce. All of these aspects enhance rural development, especially in developing countries where most of the population live in rural areas. For instance, Echeverri-Campuzano (2002) estimates that every Colombian farming family engaged in bioethanol production will earn two to three times the minimum salary (US$ 4,000/year). In South Africa meeting targets of E8 and B2 would contribute 0.11 percent to the country's Gross Domestic Product. Most of the positive effect would take place in rural areas characterized by unemployment and rising poverty (Cartwright, 2007).

Compared to other sources of energy, biofuels are labour intensive. Their production is expected to generate more employment per unit of energy than conventional fuels and more employment per unit investment than in the industrial, petrochemical or hydropower sector (UN-Energy, 2007). Creation of rural employment and the related livelihoods are all key aspects for rural development and poverty reduction. In Brazil estimations of direct employment associated with sugarcane bioethanol production ranges from 500,000 and 1 million (Worldwatch Institute, 2006; FAO, 2007) with indirect employment in the order of 6 million. Although most of them are filled by the lower-skilled, poorest workers in rural areas (Macedo, 2005), average earnings are considered better than in other sectors as the average family income of the employees ranks in the upper 50 percentile (FAO, 2007). In India, country that houses 22 percent of the world's poor, the sugarcane industry including bioethanol production is the biggest agroindustry in the country and the source of livelihood of 7.5 percent of the rural population. Half a million people are employed as skilled or semi-skilled labourers in sugarcane cultivation (Gonsalves, 2006a).

The highest impact on poverty reduction is likely to occur where sugarcane bioethanol focuses on local consumption, involving the participation and ownership of small farmers in the production and processing (FAO, 2007; Dufey *et al.*, 2007b) and where processing facilities are near to the cultivation fields.

2.5. Product diversification and value added

International sugarcane market is one of the most distorted markets. It is highly protected, in general countries manage to negotiate quotas, a limited access to different markets, and because it is a commodity, it has important price fluctuations (Murillo, 2007). In this context, sugarcane bioethanol is an opportunity to promote agricultural diversification leaving producers in a more favourable situation to deal with changes in prices and other market fluctuations. In Brazil, for instance, besides the pursue of enhanced energy security, the government promoted the PROALCOOL programme in order to deal with the fall in international sugar prices preventing thus the industry of having idle capacity (FAO, 2007). Moreover, the production of both sugar and bioethanol gives the Brazilian industry flexibility in responding to the changing profitability of sugar and bioethanol production worldwide. In most cases, sugar and bioethanol are produced in the same mills (Bolling and Suarez, 2001).

Sugarcane bioethanol can also reduce vulnerability through diversification. The changes in the European Union's sugar regime will imply that many African, Caribbean and Pacific countries will see their market access preferences eroded generating negative impacts on poverty levels. In the Caribbean, for instance, the associated possible loss of export revenues is expected to be 40 percent with a heavy contraction in the industry. The resulting sugar surpluses therefore could be accommodated for biofuels production thus helping the industry to diversify, avoiding or mitigating the expected contraction (E4Tech, 2006).

Another element to consider is the fact that sugarcane bioethanol production provides value added to sugarcane production. For instance, Murillo (2007) notes for Costa Rica that if the molasses and sugar producers substitute their production by those of bioethanol the price received would be much more than what they would get if they were to continue producing molasses or sugar for the surplus market.

2.6. Export opportunities

Although at present very little bioethanol enter the international market (about 10%), international trade is expected to expand rapidly, as the global increase in consumption (especially countries in the North) will not coincide geographically with the scaling up of production (countries in the South) (Dufey, 2006). The geographical mismatch between global supply and demand represents an opportunity for countries with significant cost advantages in sugarcane production to develop new export markets and to increase their export revenues. These are invariably developing countries in tropical and semitropical areas.

Brazil, the main global bioethanol exporter, increased its exports considerably over the last few years and today supplies about 50 percent of international demand. (Dufey *et al.*, 2007b). The Brazilian government expects that by 2015 about 20 percent of the national production to be exported (Ministerio da Agricultura *et al.*, 2006). Countries from the Caribbean Basin Initiative are developing export-oriented sugarcane bioethanol industries taking advantage of preferential market access provided by the trade agreement with the United States. Other exporters include Peru, Zimbabwe and China. As them other Latin American, African and East Asian countries are exploring the benefits of export-oriented sugarcane bioethanol sectors.

In absence of trade distorting policies and where effective distributional and social policies are supportive, the development of a successful sugarcane bioethanol export-oriented industry could effectively reduce poverty.

2.7. Reduced greenhouse gas emissions

At present global warming is considered one of the key global threats facing the humanity (Stern, 2006). Biofuels alleged reduced greenhouse gas emissions compared to fossil fuels are one of the main policy rationales for their promotion especially in Northern countries. There are two ways in which biofuels can reduce carbon emissions. First, over their life cycle, biofuels absorb and release carbon from the atmospheric pool without adding to the overall pool (in contrast to fossil fuels). Second, they displace use of fossil fuels (Kartha, 2006). However, biofuels production does, in most cases, involve consumption of fossil fuels.

Compared to other types of liquid biofuels and under certain circumstances, Brazilian sugarcane bioethanol and second generation biofuels show the higher reductions in greenhouse gas emissions relative to standard fuels. IEA (2004) estimates that greenhouse emissions from sugarcane bioethanol in Brazil are 92 percent lower than standard fuel, while wheat bioethanol points to reductions ranging from 19 percent to 47 percent and reductions from sugar beet bioethanol vary between 35 percent and 53 percent. In addition to Brazil's exceptional natural conditions in terms of high soil productivity and that most sugarcane crops are rain fed, a key factor behind its great greenhouse emissions performance is that nearly all conversion plants' processing energy is provided by 'bagasse' (the remains of the crushed cane after the juice has been extracted). This means energy needs from fossil fuel are zero and the surplus bagasse is even used for electricity co-generation. In 2003, Brazil avoided 5.7 million tonnes CO_2 equivalent due to the use of bagasse in sugar production (Macedo, 2005). Moreover, new developments in the sector such as the commercial application of lignocelulosic technology that will allow the use of bagasse for bioethanol production and the increased generation of electricity from bagasse will improve their greenhouse emissions balance (Dufey *et al.*, 2007a).

However the Brazilian experience is not necessarily replicable in other contexts. For example, efficiency gains and the greenhouse emissions reductions associated with co-generation are an option for those countries whose electricity sectors regulation allows power sale to the grid (E4Tech, 2006).

Finally, these estimations do not include the emissions resulting from changes in land use and land cover induced by sugarcane plantations for bioethanol production. For example, the evaluation of greenhouse emissions from Brazil for the 1990-1994 period points out the change in land use and forests as the factor accounting for most of the emissions (75%), followed by energy (23%). This implies that if additional land use for sugarcane production leads (directly or indirectly) to conversion of pastures or forests as suggested later in this chapter, the greenhouse emissions may be severe and could have a major impact on the overall greenhouse emission balance (Smeets *et al.*, 2006). Overall, the land use issue requires further attention and is addressed in another chapter of this book.

2.8. Outdoor air quality

Road transport is a growing contributor to urban air pollution in many developing country cities. One of the greatest costs of air pollution is the increased incidence of illness and premature death that result from human exposure to elevated levels of harmful pollutants. The most important urban air pollutants to control in developing countries are lead, fine particulate matter, and, in some cities, ozone. Sugarcane bioethanol, when used neat, is a clean fuel (aside from increased acetaldehyde emissions). More typical use of bioethanol is in low blends. Bioethanol also has the advantage of having a high blending octane number, thereby reducing the need for other high-octane blending components such as lead that cause adverse environmental effects. Venezuela, for instance, began importing Brazilian bioethanol as part of the effort to eliminate lead from gasoline. Bioethanol can be effective for cutting carbon monoxide emissions in winter in old technology vehicles as well as hydrocarbons emissions. The latter are ozone-precursors, in old technology vehicles (ESMAP, 2005b).

On the other hand, there is air pollution associated with the slush and burn of sugarcane and the burning of the straw, a common practice in developing countries to facilitate the harvesting. This issue is further addressed in Section 3.b on Environmental Impacts.

2.9. Opportunities for investment attraction – including the Clean Development Mechanism

Developing countries can make use of the carbon finance markets for attracting investment into biofuels projects using the market value of expected greenhouse emission reductions. The Clean Development Mechanism (CDM) under the Kyoto Protocol is the most important example of the carbon market for developing countries. The CDM allows developed countries (or their nationals) to implement project activities that reduce emissions in developing countries in return for certified emission reductions (CERs). Developed countries can use the CERs generated by such project activities to help meet their emissions targets under the Kyoto Protocol. For instance, it is calculated the Colombian Programme on bioethanol would reduce CO_2 emissions by six million tons, offering opportunities to obtain financial resources for the project trough the CDM (Echeverri-Campuzano, 2000). For Costa Rica, Horta (2006) estimates that considering an avoided ton of carbon at a conservative price of US$ 5, in the scope of the Kyoto Protocol and the valid mechanisms of carbon trade, US$ 320,000/year can be obtained using a 10 percent of sugarcane bioethanol in the gasoline blend.

Although the CDM is a potential source of financing for biofuels projects, taking advantage of it can present a number of challenges for the developing country host. Firstly, so far there is no liquid-biofuels baseline and monitoring methodology approved. Calculation of greenhouse gases emissions is not straightforward and for many countries biofuels are still a relatively expensive means of reducing these emissions relative to other mitigation measures. An additional challenge is that the existing experience with CDM projects shows that approved projects are strongly concentrated in a handful of large developing countries, with over 60 percent of all CDM projects distributed across China, India and Brazil alone. While there are simplified procedures for small-scale projects, the current structure of the CDM tends to select for large-scale projects. The transaction costs associated with registering a CDM project are often prohibitively expensive for smaller developing countries, which imply that economies of scale are relevant (Bakker, 2006). For bioenergy projects specifically, the exclusion of all land use activities from the CDM except for afforestation and reforestation is another significant limiting factor, since in the poorest developing countries, land-use related emissions make up the bulk of greenhouse gases emissions from biomass energy systems (Schlamadinger and Jürgens, 2004). Overall, as FAO (2007) concludes, while carbon credits might be influential in the future, currently the carbon market does not have a large influence over the economics of bioenergy production.

3. Risks and challenges

Section 2 analysed a diverse range of benefits associated with sugarcane bioethanol in terms of its potential to support poverty reduction and environmental sustainability. However, as this section argues, these benefits are not straightforward. There is a range of challenges and trade-offs that need to be confronted in order to realize the full potential that sugarcane bioethanol offers to support the MDGs, which include: (a) impacts on food security; (b) environmental pressure; (c) small farmer inclusion and fair distribution of the value chain benefits; (d) land impacts; (e) employment quality; (f) need of government support; (g) existence of market access and market entry barriers and; (h) issues related to improved efficiency, access to technology, credit and infrastructure. These issues are addressed in the following.

3.1. The food versus fuel debate

Current food prices increases, the role that biofuels play on such rises and their related impacts on food security are, probably, one of the most controversial debates being held both at national and international fora. Indeed, food prices increased by 83 percent during the last three years (World Bank, 2008). The Food and Agriculture Organization of the United Nations (FAO) food index price rose by nearly 40 percent in 2007, from a 9 percent increase in 2006 (IFPRI, 2008). World prices rose much more strongly in 2006 than anticipated for cereals, and to a lesser extent for oilseeds, but weakened for sugar (OECD-FAO, 2007).

The understanding of biofuels impacts on food security is a wider and complex. It requires considering that the link between food prices increases and food security is not unique and necessarily negative. It needs to be analysed in the context that changes in food prices not only impact food *availability* but also its *accessibility* through changes in incomes for farmers and rural areas (Schmidhuber, 2007).

3.1.1. Impacts on food availability

The key question at the national level is whether the savings and gains from biofuels will outweigh additional food costs. Biofuels compete with food crops for land and water, potentially reducing food production where new agricultural land or water for irrigation are scarce (Dufey *et al.*, 2007b). For biofuels that are manufactured from food crops, there is also direct competition for end-use. To what extent sugarcane bioethanol creates competition for land and crowd out food crops is an issue that is not very clear. The limited available evidence would suggest a lesser impact compared to other feedstocks. Zarrilli (2006), for example, points out that sugarcane producing regions in Brazil stimulate rather than compete with food crops, which is done by two means. Firstly, through the additional income generated by sugarcane related agro-industrial activities which 'capitalises' agriculture and improves the general conditions for producing other crops. This is also noted by Murillo (2007) for Costa Rica, where under current weather conditions and land use, sugarcane bioethanol production is seen as a complement in income generation rather than a competition for basic products and vegetables. Secondly, the high productivity of cane per unit of land compared to other feedstocks enables a significant production of cane, with a relatively small land occupation (Zarrilli, 2006). Sugarcane's minimal land requirements but in the context of sub-Saharan Africa is noted by Johnson *et al.* (2006), but needs to be proven (Dufey *et al.*, 2007b). Moreover, in those countries where bioethanol is produced from sugarcane molasses there is no displacement of food crops (Rafi Khan *et al.*, 2007). In addition, in many African countries, cassava and maize are grown for subsistence purposes while cane is often grown for sugar export. Diversion to fuel production is therefore more likely to adversely affect food availability in the case of cassava (Johnson and Rosillo-Calle, 2007)

At the international level, the growing international demand for biofuels is expected to reverse the long-term downward trend in global prices of agricultural commodities. Several studies have been conducted linking increased global biofuels production with rising agricultural commodity prices. Estimations vary widely with most credible ones going up to 30 percent. Other contributing factors to price increases are the weather-related shortfalls in many key producing countries, reduced global stocks, increased demand from new emerging economies in Asia (OECD-FAO, 2007) and speculation (IFPRI, 2008). In that sense, the higher demand for biofuel feedstocks is viewed as increasing pressure on an already tight supply.

However, it is one issue trying to isolate how much biofuels, in overall, are responsible for the sector's inflationary pressure and, a different one, understanding to what extent sugarcane bioethanol is responsible for the price increase. Although the available evidence in this sense is also scant, it would suggest that, compared to other feedstocks, sugarcane bioethanol would have a slighter impact on food security. A key reason behind this is that sugarcane is not a principal food crop. Staple grains like maize and rice are often the main food source for the poorest people, accounting for 63 percent of the calories consumed in low-income Asian countries, nearly 50 percent in Sub-Saharan Africa, and 43 percent in lower-income Latin American countries (IFPRI, 2008). Rosegrant (2008) in an exercise in which biofuel production was frozen at 2007 levels for all countries and for all crops used as feedstocks, shows the smaller price reductions for sugarcane followed by wheat while the higher reductions are for maize (Figure 1). Another reason been argued is that sugarcane price would be relatively uncorrelated with other food crops (Oxfam, 2008).

3.1.2. Impacts on accessibility

The issue of how the gains and costs of biofuels to food security are distributed across society has been less explored in the literature. FAO and other commentators agree that hunger is largely a matter of access rather than supply, so that a focus on rural development and livelihoods makes more sense that trying to maximise global food supply, which for now at least is adequate for global needs (Murphy, 2007).

Higher agricultural commodity prices are good news for agricultural producers, but they have an adverse impact on poorer consumers, who spends a much larger share of their income on food (IFPRI, 2008). There are also differences depending on whether households are net food producers or buyers. For small farmers that are net food producers, overall gains in welfare and food security are expected due to rising revenues from biofuel crops and

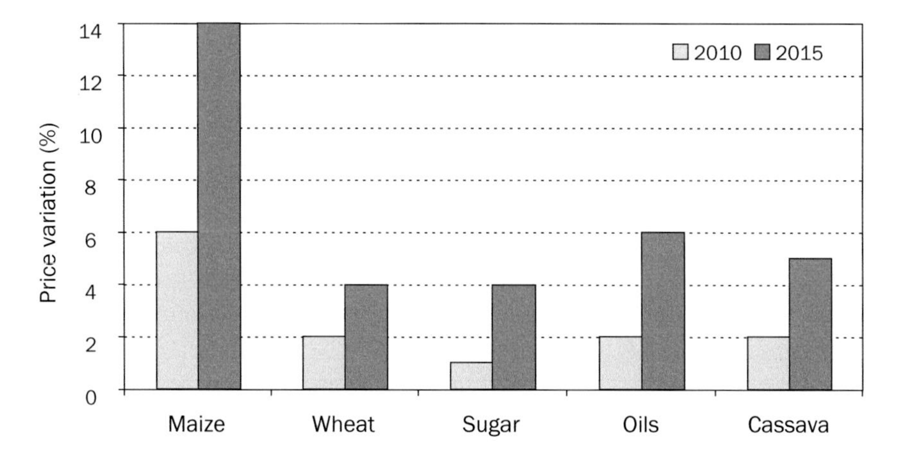

Figure 1. Change in selected crop prices if biofuel demand was fixed at 2007 levels. Source: Rosegrant (2008).

food crops (Peskett *et al.*, 2007). In overall, poor consumers in urban areas who purchase all their food are expected to be worst off. From this perspective and compared to other feedstocks, sugarcane bioethanol is likely to provide more limited opportunities to meet food security for small farmers. In Brazil, for example, sugarcane is a crop mainly grown under large-scale schemes, with limited participation of small farmers. In regions such as Asia, although small farmers participation in sugarcane cultivation is important, the need to use irrigation makes more unlikely to involve poorest farmers (ICRISAT, 2007). More widely, it is agreed that despite being producers of agricultural crops, most poor farming households in rural areas are net buyers of food (Dufey *et al.*, 2007b; IFPRI, 2008).

Finally, it should be noted that, historically, domestic food prices have not been tightly linked to international food or energy prices, as price transmission mechanisms are not straightforward (Hazell *et al.*, 2005). For instance, agricultural pricing policies such as price fixation, the remoteness of some rural areas, trade distortions and power structures governing agricultural commodity markets are key factors preventing world prices from reaching domestic markets. This may imply that farmers may not see the incentives to change feedstock production in tandem with changes in international prices.

3.2. Environmental pressure

Traditional environmental impacts associated with sugarcane appear when it comes to managing soil, water, agrochemicals, agricultural frontier expansion and the related biodiversity impacts. Among them, impacts on agricultural frontier and on water deserve especial attention. Regarding the former, it should be noted that the bulk of the sugarcane expansion in the last thirty years in Brazil has been concentrated in the central southern region of the country. Between 1992 and 2003, 94 percent of the expansion occurred in existing areas of agriculture or pastureland and only a small proportion of new agricultural borders were involved (Macedo, 2005). Often the sugarcane crop replaced cattle grazing and other agricultural activities (e.g. citrus crops), which in turn moved to the central region of Brazil where the land is cheaper (Smeets *et al.*, 2006). Land converted to agriculture in the sensitive area of the Cerrado savanna (which accounts for 25% of the national territory) has been used for cattle grazing and/or planted to soya, with only a small proportion for sugarcane. However, given the new phase of expansion experiencing the sector for bioethanol production, new areas are expected to be converted to sugarcane, including the Cerrado of Mato Grosso do Sul, Goiás and Minas Gerais (Dufey *et al.*, 2007a). This could further increase the pressure on the already affected biodiversity and produce greenhouse emissions. There is concern in this sense on the impacts that the substitution effect - sugarcane taking over existing pastureland or other crops that become less profitable which in turn advance into protected or marginal areas – may have on biodiversity. Indeed, in Brazil, substitution effect related impacts are considered more significant than the direct effects of sugarcane expansion (Dufey, 2007). In Africa, on the other hand, land constraints appear unlikely in any near-term scenario, and resources such as water, as explained in the next paragraph, may turn out to be the key limiting factor (Johnson and Rosillo-Calle, 2007).

Regarding water, sugarcane requires large amounts of water, both at the farming and processing level. Even in Brazil where most sugarcane is rain fed, irrigation is increasing. Energy cane, which is especially bred for energy production, requires more water and fertiliser than conventional sugarcane (Cloin, 2007). Water is likely to be a key limiting factor especially in dry and semi-dry areas in Africa and Asia. Bioethanol impact on water quality is another issue and not only at the farming level due to the use of agrochemicals but also at the processing level. Vinasse, - a black residue resulting from the distillation of cane syrup - is hot and requires cooling. In the mountainous areas of north-eastern Brazil, for instance, the costs of pumping storing vinasse were prohibitive, and it was therefore released into rivers, resulting in the pollution of rivers causing eutrophication and fish kills. Currently, vinasse is used for ferti-irrigation of cane crops, together with wastewaters. Moreover, legislation has been implemented in Brazil to avoid the negative impacts of vinasse applications, although its coverage is incomplete and its enforcement is rather weak (Smeets *et al.*, 2006). All in all, while steps have been taken in Brazil order to manage vinasse disposal, in countries such as Malawi it is still a major concern (Johnson and Rosillo-Calle, 2007).

Furthermore, the air pollution associated with the slush and burn of sugarcane and the burning of the straw, a common practice in developing countries to facilitate the harvesting, is an additional issue. Sugarcane burning emits several gases including CO, CH_2, ozone, non-methane organic compounds and particle matter that are potentially damaging for human health. Several studies were conducted in São Paulo in Brazil during the 1980s and 1990s to identify the impacts of sugarcane burning on human health. Although some studies did not found a link, others studies did confirm the relationship (Smeets *et al.*, 2006;

Dufey *et al.*, 2007a). Legislation has been passed in Brazil by which sugarcane burning is to be completely phased out in the São Paulo State by 2031. In Southern Africa efforts to reduce sugarcane burning pre-harvesting have also been reported (Jackson, 2004), but in other countries it still remain a major practice.

Overall, sugarcane bioethanol production poses some specific environmental challenges that need to be carefully identified and managed using a life cycle approach in order to achieve the MDG on environmental sustainability.

3.3. Small farmers inclusion and fair distribution of the value chain benefits

Addressing poverty means that biofuels should benefit poor and small farmers overall. An emphasis on small farmers would provide livelihoods across the greatest section of the populations (Johnson and Rosillo-Calle, 2007). But the competitiveness of a biofuels industry is highly dependent on gaining economies of scale. Often large-scale systems are more globally competitive and export oriented, while small-scale systems offer greater opportunities for employment generation and poverty alleviation (Dufey *et al.*, 2007b). In Brazil, the sugarcane business model is characterised by enormous concentration of land and capital, which highlights the need for a better inclusion of small-scale producers (Dufey *et al.*, 2007a). Increasing economies of scale and land concentration have meant that benefits of sugarcane bioethanol production for small land owners have so far been limited and large farmers and industrialists have benefited more from the expansion of the industry (Peskett *et al.*, 2007). In contrast, in countries such as India and South Africa small farmers are key players in the sugarcane sector. In India, they represent between 60 and 70 percent of the cane growers (Johnson and Rosillo-Calle, 2007). In Costa Rica, the proportion of small producers in the sugarcane sector increased by 97 percent between 2000 and 2005 (Murillo, 2007).

Small farmers face several obstacles in trying to access supply chains. They trade-off high transportation costs getting crops to processing plants with selling through middlemen (Peskett *et al.*, 2007; Rafi Khan *et al.*, 2007). In India, farmers must access to irrigation to be competitive, which is increasingly difficult and expensive due to growing water scarcity and cost (ICRISAT, 2007). At processing plants they have to time delivery to fit daily plant capacity and meet plant standards. Either way, small producers are price-takers (Peskett *et al.*, 2007). Box 3 highlights some of the challenges faced by sugarcane small farmers in Pakistan.

However, large-scale and small-scale systems are not mutually exclusive and can interact successfully in a number of different ways (Dufey *et al.*, 2007b). Some of the models for partnership between large-scale and small-scale enterprises include outgrower schemes, cooperatives, marketing associations, service contracts, joint ventures and share-holding by small-scale producers (Mayers and Vermeulen, 2002). Concerning sugarcane, in Brazil co-operatives operate in certain areas (Oxfam, 2008). In India some of the sugar mills are

Box 3. Unfair distribution of benefits against small farmers - middleman in Pakistan.

In Pakistan, where bioethanol is produced from sugarcane molasses, middlemen play a key role in sugarcane procurement and often end up exploiting small-scale farmers forcing them to sell at distress prices. In collusion with mill owners, they orchestrate delays at the mill gate; the problem becomes exacerbated during surplus years. The farmer has no option but to accept the price offered (lower than the support price) or face further delays. Large farmers are better placed as their crop represents a large proportion of the mill intake and they also have greater political clout. Small farmers are indebted to middlemen for their consumption and input needs, which also leads to under pricing. Further, a report by the Agricultural Prices Commission of Pakistan indicates that the scales installed to weigh sugarcane do not provide correct readings. However, given the high level of illiteracy among small-scale growers, such practices go undetected. Moreover, mills are also known to make undue deductions contending that sugarcane quality is low and contains high trash content.

Source: adapted from Rafi Khan *et al.* (2007).

cooperatives in which farmers also hold ownership shares in the factory (ICRISAT, 2007). The South African sugar industry distinguishes itself by operating a successful small-scale outgrower scheme, which supplies 11 percent of the country's sugarcane under contract farming arrangements to one of the three major mills (Cartwright, 2007).

The need for economies of scale to increase competitiveness constitutes a pressure to reduce costs. The main mechanisms for doing this – introduction of improved varieties, switch away from diversified production systems to monocropping, move to larger land holdings, and shift to increasingly capitalised production - are difficult or risky for small producers. For example, in Brazil, selection of improved cane varieties (e.g. energy cane) and investment in irrigation have helped to improve yields but the benefits of these have mostly been felt on plantations. Other mechanisms, such as increasing labour productivity without increasing wages, are likely to be detrimental to poor households (Peskett *et al.*, 2007). This presents a serious challenge to identifying pro-poor biofuels production systems.

Analysis by a UN consortium suggests that efficient clusters of small and medium-scale enterprises could participate effectively in different stages of the value chain (UN-Energy, 2007). The main challenge is how to provide appropriate policy conditions to promote value-sharing and prevent monopolisation along the chain (Dufey *et al.*, 2007b). Controlling value-added parts of the production chain 'is critical for realising the rural development benefits and full economic multiplier effects associated with bioenergy' (UN-Energy, 2007). In countries such as Thailand policy interventions are addressing the sharing of the earning between sugarcane growers and producers (70% and 30%, respectively). However, for bioethanol manufactured directly from sugarcane juice, producers argue the Government has to come with a better agreement as they have to invest on bioethanol plants (Gonsalves, 2006b).

At the international level this implies that the biofuels value chain must shift to the countries that produce the feedstock.

Overall, economies of scale are important and small-farmers will need to adapt and get organised towards that direction. Challenges and difficulties will be confronted and more research is needed to understand the role partnership schemes (Dufey *et al.*, 2007b).

3.4. Landlessness and land rights

The strength and nature of land rights are key determinants of patterns of land ownership under biofuel production. As the above point suggests, the need of costs reduction offers considerable incentives for large-scale, mechanised agribusiness and concentrated land ownership. This is turn can displace small farmers and other people living from the forests and depriving them from its main source of livelihoods. This may have devastating effects on rural poverty. Indeed, the primary threat associated with biofuels is landlessness and resultant deprivation and social upheaval, as has been seen for example with the expansion of the sugarcane industry in Brazil (Worldwatch Institute, 2006; Dufey *et al.*, 2007b) which is summarised in Box 4. Johnson and Rosillo-Calle (2007) also highlight land related problems in the African context, where the high proportion of subsistence farming and complexities of land ownership under traditional land regimes make large acquisition of land, for large-scale sugarcane operations, a highly controversial issue.

Box 4. Access, ownership and use of land in Brazil.

Biothanol production in Brazil has inherited problems faced by the sugar industry over the last 50 years, including violent conflict over land between indigenous groups and large farmers. Problems stem from weak legal structures governing land ownership and use which have increased land concentration, monoculture cropping and minimisation of production costs. Land occupation planning is carried out at municipal level, but not all municipalities have developed guidelines governing monocultures. Land concentration in Brazil is very high, with only 1.7% of real estate covering 43.8% of the area registered. Land concentration and subsequent inequality is increasing with expansion of monocropping areas, reduction of sugar mill numbers, growth in foreign investment and land acquisition. The need of economies of scale for efficient sugarcane production in part drives these effects.

Source: adapted from Peskett *et al.* (2007).

Rossi and Lambrou (2008) note some gender-differentiated risks. Marginal lands are particularly important for women. The conversion of these lands to energy crops might cause displacement of women's agricultural activities towards increasingly marginal lands, with negative effects in their ability to meet household obligations. This highlights the urgent need of a careful analysis of what the concept of 'marginal', 'idle' or 'unproductive' lands really entails. It is in these lands where most government are mandating biofuels to be grown.

3.5. Quality of the employment

Sugarcane bioethanol will generate a range of employment opportunities, mostly in rural areas, which is certainly good for poverty reduction. However there are limitations and trade-offs. Firstly, there is concern about the quality of employment, whether self-employment (small-scale farmers) or employment within large-scale operations (Worldwatch Institute, 2006; UN-Energy, 2007). Sugarcane harvesting is extreme physically demanding. Production is highly seasonal and, in Brazil, for example, the ratio between temporary and permanent workers is increasing. Low skilled labour dominates the industry and a high rate of migrant labour is employed. In southern Africa the sudden influx of seasonal workers has had negative effects on community cohesion, causing ethnic tension and disintegration of traditional structures of authorities. Migrants behaviour is also linked with higher rates of HIV infection around sugarcane plantations (Johnson and Rosillo-Calle, 2007).

Whilst over the latest years in some plantations in Brazil improvements in working conditions have been done, in other plantations, sugarcane cutters continue to work in appalling conditions. Cases of forced labour and poor working conditions within the sector are still reported (Oxfam, 2008). Other problems include a lack of agreed or enforceable working standards in many countries, and lack of labour representation (Dufey *et al.*, 2007b).

Moreover, compared to other feedstocks (e.g. palm oil, castor oil, sweet sorghum) sugarcane is less labour-intensive and thus provide less on-farm and off-farm employment (Dufey *et al.*, 2007b). The industry greater mechanisation in turn reduces labour demands. One harvester can replace 80 cutters and thus facilitate the whole harvesting process (Johnson and Rosillo-Calle, 2007). In Brazil mechanization of sugarcane harvesting has been driven by increasing labour costs and more recently by legislation to eliminate sugarcane burning. Total employment in the industry decreased by a third between 1992 and 2003 (ESMAP, 2005b). Indeed sugarcane related unemployment is expected to become the key social challenge faced by the sugarcane industry in Brazil (Dufey *et al.*, 2007a). This can have devastating effects on poverty levels as it is unemployment among the lower-skilled workers.

In order to balance trade-offs between environmental needs, mechanisation and unemployment, Johnson and Rosillo-Calle (2007) propose the use of half-mechanisation which was successfully used in Brazil as a transition towards full mechanisation. It consists in mechanical aid for the harvesting, in which a machine is used for cutting the cane and workers are used to gather the crops. As the cutting of the cane is the hardest part physically, the authors argue this system would also contribute to opening up the labour force for women.

All in all, although recognising that many of the above mentioned issues are not exclusive for sugarcane bioethanol, employment generation that leads to effective poverty reduction requires addressing these problems.

3.6. Government support

Experience suggests the biofuels sector requires some form of policy support, at the very least in the initial phases development. Even Brazil, the most efficient biofuel producing country, still maintains a significant tax differential between gasoline and hydrous ethanol to promote the sector (ESMAP, 2005b) and fixes a mandatory blend (between 20% to 25%). More generally, the PROALCOOL programme in the past required heavy support. Between 1975 and 1987 it produced savings for US$ 10.4 billion but it costs were US$ 9 billion (World Watch Institute, 2006). Moreover, with falling oil prices, rising sugar prices, and a national economic crisis the programme simply became too expensive and collapsed by end of 1980s.

In many countries, the main rationale behind biofuels production is to decrease the costs associated with imported fossil fuels. Among the costs of such a policy that need to be accounted is the foregone duty on fuel imports, which results in a decline in government revenues. For instance, in Brazil, the forgone tax revenue in the state of São Paulo, which accounts for more than one-half of the total hydrous ethanol consumption in the country, was about US$ 0.6 billion in 2005 (ESMAP, 2005b). In many developing countries a substantial portion of public revenues are derived from import duties. In addition, the diversion of sugar exports for bioethanol production for domestic markets means that countries may suffer reductions in their export earnings. All these pose significant challenges in poorest countries, where there are a multitude of urgent needs competing for scarce fiscal resources.

Another issue is that once granted and the biofuel industry has been launched, subsidies are difficult to withdraw. A major challenge to reduce policy support is the vested interests created in the domestic industry (Henniges and Zeddies, 2006).

On the other hand, the existence of contentious domestic policies and practices can undermine industry development. For instance, Rafi Khan *et al.* (2007) and Gonsalves (2006a) report the negative effects on bioethanol development of policy measures such as a high central excise duty and sales tax on alcohol that exist in Pakistan and India, respectively. The lack of policy provenance - reflected by the fact that the Pakistani government directed the Petroleum Ministry (who houses the oil lobby) to develop the bioethanol conversion plan also constitutes an additional policy constraint. Pricing issues - whether to use bioethanol international price or its cost of production - can also affect industry development (Rafi Khan *et al.*, 2007).

All the above suggest the promotion of a sugarcane bioethanol industry can become very expensive, not only due to the high up front investments that are required but also due to the financial resources that are needed to make it viable in the long term.

From a poverty reduction strategy point of view this means that governments should design their sugarcane bioethanol policies so as to reach the desired target group. As ESMAP (2005b) notes, resources that flow to agriculture all too often benefit politically powerful, large producers and modern enterprises disproportionately at the expense not only of the society as a whole, but of those that are supposed to be the main beneficiary group: smallholder farmers and landless workers. Examples include untargeted producer subsidies and distortionary subsidies for privately used inputs such as water and electricity. According to the same source, promoting biofuels for energy diversification can make sense if large government subsidies are not required. However, UN-Energy (2007) holds the view that if the large subsidies are targeting small producers this may be money well spent. Governments tend to get higher returns on their public spending by fostering small-scale production due to the lowered demand for social welfare spending and greater economic multiplier effects.

Overall, governments need to conduct a careful assessment of the pros and cons of promoting sugarcane bioethanol to support poor rural communities versus those of other alternatives. Similarly, from a climate change mitigation strategy, although sugarcane bioethanol may show the greatest greenhouse reductions compared to other first generation feedstocks, these should be assessed against the costs of other policy instruments to achieve the same goal.

3.7. Market access and market entry barriers

The strategic nature of bioethanol implies the existence of some degree of protectionism in almost any producing country. Protectionism is especially acute where energy security is equated with self-sufficiency or where biofuels are promoted to help domestic farmers in high-cost producing countries (Dufey *et al.*, 2007b). The use of tariffs to protect domestic biofuel industries is a common practice and, as Table 1 shows, these can be very high. However, these tariffs are only indicative as their actual level applied vary widely as both the European Union and the United States have trade agreements providing preferential market access to several developing countries. In particular, the extra US$ 0.14 to each litre (US$ 0.54 per gallon) of imported bioethanol on top of the 2.5 percent tariff applied by the United States, it is said to be targeting Brazilian imports as it brings the cost of Brazilian bioethanol in line with that produced domestically (Severinghaus, 2005). Tariff escalation, which discriminates against the final product, can also be an issue, for example, where there are differentiated tariffs on bioethanol and feedstock such as raw molasses (Dufey, 2006).

Table 1. Import tariffs on bioethanol[1].

Country	Import tariff
US	2.5% + extra US$ 14 cents/litre (46% *ad valorem*)
EU	€ 19.2/hl (63% *ad valorem*)
Canada	4.92 US$ cent/litre
Brazil[2]	20%
Argentina	20%
China	30%
Thailand	30%
India	186% on undenatureated alcohol

Source: adapted from Dufey *et al.* (2007b)
[1] Undenaturated alcohol.
[2] Temporarily lifted in February 2006.

On the other hand, the planning of an export-oriented bioethanol industry based on the rationale of preferential market access is a risky strategy. As Box 5 suggests for Pakistan, trade preferences can be withdrawn at any time with devastating effects on the industry.

Subsidies is another key concern. In industrialised countries, government support for the domestic production of energy crops, the processing or commercialisation of biofuels seems to be the rule (Dufey, 2006). Amounts involved are enormous. In the United States, Koplow (2006) estimated that subsidies to the biofuels industry to be between US$ 5.5 billion and US$ 7.3 billion a year. In the European Union, Kutas and Lindberg (2007) estimated that total support to bioethanol amounted € 0.52/litre.

The impacts these policies have on the developing countries competitiveness and on their potential for poverty reduction needs to be understood as domestic support in these countries is likely to be very limited. Moreover, subsidies impacts on environmental sustainability are also questionable as they promote bioethanol industries based on the less efficient energy crops and with the least greenhouse gases reductions such as maize and wheat (Dufey, 2006).

The proliferation of different technical, environmental and social standards and regulations for biofuels – without a system for mutual recognition – cause additional difficulties. For instance, at present not all biofuels are perceived as 'sustainable' especially those coming from overseas. As a consequence, several initiatives towards the development of sustainability certification for both bioethanol and biodiesel have started. Some of them are led by governments (e.g. the United Kingdom, Netherlands and the European Union); others by

Box 5. The elimination of Pakistan from the EU GSP.

Until recently, Pakistan was the second largest industrial alcohol exporter to the EU after Brazil, under the General System of Preferences (GSP). In May 2005, the Commission of Industrial Ethanol Producers of the EU (CIEP) accused Pakistan and Guatemala (the largest duty free exporters for the period 2002-2004) of dumping ethyl alcohol in the EU market, causing material harm to domestic producers. The Commission dropped proceedings a year later when full custom tariffs were restored on Pakistani imports. Later, following a complaint lodged by India at the World Trade Organization (WTO), a panel concluded that by granting tariff preferences to 12 countries under this special arrangement the EU was violating GATT/WTO preferential treatment obligations. The EU consequently removed Pakistan from the GSP. In the revised GSP regime, the anti-drug system has been replaced by GSP Plus, for which Pakistan does not qualify.
Elimination of Pakistan from the GSP had devastating effects on the local industry. Distilleries begun to suffer important losses and some had no option but to cease operations. Whilst between 2002 and 2003, the number of distilleries in the country increased from 6 to 21, the more stringent EU tariff measures together with a rise in molasses exports, the distilleries were soon running idle capacities. Currently, at least 2 distilleries have shut down, with another 5 contemplating that option.

Source: adapted from Rafi Khan *et al.* (2007).

NGOs (e.g. WWF); and also by Universities (e.g. Lausanne University). These schemes tend to focus on traditional environmental and social aspects of feedstocks production, with several of them including greenhouse emission issues and with some few of them expanding to food security concerns. Although environmental and social assurance is needed in the industry, where these schemes are developed by importing nations, with little participation by producing country stakeholders, insufficient reflection of the producing countries' environmental and social priorities and without mutual recognition between them, they are bound to constitute significant trade barriers. Moreover, the experience with assurance schemes in the agriculture and forestry sector indicates that the complex procedures and high costs usually associated with them have regressive effects in detriment of small and poorest producers in developing countries. All in all, sustainability standards for bioethanol trade are to become more and more important. Countries wanting to benefit from bioethanol exports need to invest in the development of robust and credible certification systems that satisfy importing countries requirements.

Overall, it is widely agreed that developing countries would benefit from enhanced bioethanol trade and therefore the need to eliminate trade barriers.

3.8. Improving efficiency, access to technology, credit and channelling investment

The development of a successful bioethanol sector goes beyond having available land, cheap labour and good climate. It crucially depends on countries' domestic capacity to expand production efficiently, accessing the technology and assuring best practice. Indeed, Brazil's success in developing an efficient bioethanol industry is in a large extent explained by the enormous endogenous efforts devoted to R&D, capacities building and infrastructure (Dufey *et al.*, 2007a). This implies that having a number of technical skills for research, technology transfer as well as access to credit are critical issues. Moreover, those countries wanting to develop an export oriented sector also need to be in compliance with the relevant technical standards in importing markets and to invest in suitable transport infrastructure (roads, water ways and ports) to reach exports markets. Countries also need to have sufficient capacity in policy implementation and project management to run biofuels production and processing effectively (Dufey *et al.*, 2007b).

At present, many countries foresee a major participation of the sugar industry in bioenergy production. However, the current low efficiency and productivity of the sector in many of them implies that major changes to the industry's structure will be needed to make sugarcane an important feedstock (FAO, 2007). In countries where bioethanol is produced from molasses and wanting a significant scale of production, efforts will need to be made to produce from sugarcane juice, which is a relatively more efficient source of bioethanol and capable of supplying larger volumes (Woods and Read, 2005). Other specific needs include adaptive agricultural research and extension development for enhanced transfer of bioethanol technologies. Investment is also important to bring agricultural practices up to the required level of technical capacity, scale of operations, and intensity of production (Johnson and Rosillo-Calle, 2007)

4. Conclusions

Sugarcane bioethanol can contribute to the achievement of several Millennium Development Goals through a varied range of environmental, social and economic advantages over fossil fuels. The highest impact on poverty reduction is likely to occur where sugarcane bioethanol production focuses on local consumption, involving the participation and ownership of small farmers and where processing facilities are near to the cultivation fields.

Realising the greatest potential of sugarcane bioethanol on poverty reduction implies that several challenges will need to be confronted and dealing with serious trade-offs. Especially tough will be those related to efficiency gains through large-scale operations, mechanisation and land concentration versus small farmers inclusion. Economies of scale are important and small farmers will need to adapt and get organised towards that direction. Likewise, the resulting unemployment among the lower-skilled workers is a key aspect to be addressed. Whilst the domestic use of sugarcane bioethanol may imply opportunities in terms of general well-being, the increasing use of marginal land for biofuels cultivation may imply negative impacts among the most vulnerable such as women. From a poverty reduction strategy this means that governments should explicitly design their sugarcane bioethanol policies to provide the right environment to promote business models that maximises rural development, small farmer inclusion and equitable access to ownership and value along the chain. One example in that direction can be the use of tax-breaks for companies that include small producers among their suppliers, which is already being used in the context of biodiesel in Brazil through the PROBIODIESEL programme.

The impacts of sugarcane bioethanol on food security are less clear. Regarding food availability and compared to other feedstocks, sugarcane bioethanol would provide better opportunities to meet food security as long as it creates less competition for land and crowd out other crops. However, from an accessibility point of view, it would provide more limited opportunities to the extent that its production is less likely to involve small or poorest farmers. Overall, more research is needed to understand these linkages.

From an environmental sustainability perspective, compared to other first generation biofuels, sugarcane bioethanol offers opportunities to achieve one of the greatest reductions in greenhouse emissions under certain circumstances. However, available estimations need to be revised to include the emissions directly and indirectly associated with changes in land use and cover. Similarly, biodiversity impacts linked to changes in land use and cover especially those associated with the substitution effect appear as crucial environmental aspects to be addressed and more research to understand them is needed. Likewise, impacts on water, especially in the context of dry and semi-dry lands, are other key aspects that deserve better analysis. Only the adequate understanding and management of these impacts, using a life cycle approach, will help to improve the environmental sustainability of sugarcane bioethanol and thus achieving the Millennium Development Goal on environmental sustainability.

In some contexts, the promotion of a sugarcane bioethanol industry can be a very expensive means of achieving poverty reduction and promoting environmental sustainability. Governments need to conduct a careful assessment of the pros and cons of promoting sugarcane bioethanol to support poor rural communities versus those of other policy choices. Similarly, from a climate change mitigation strategy, although under certain circumstances sugarcane bioethanol shows the greatest greenhouse reductions compared to other first generation feedstocks, these should be assessed against the costs and benefits of other policy instruments for achieving the same goal.

Another crucial issue involved in realising the full potential of sugarcane bioethanol is the building of an adequate set of national capabilities on technical skills, policy implementation, project management and development of R&D programmes. These should come hand in hand with promoting access to technology, credit and finance as well as the provision of some minimum transport infrastructure. For those countries wanting to take advantages of an export oriented industry, capacities building on standard setting and compliance as well as the negotiation of favourable terms of trade constitute other key aspects.

Policy coherence is another issue. The promotion of a sugarcane bioethanol sector that contributes to sustainable development and poverty reduction should be aligned with existing relevant national and international policies and frameworks such as Sustainable Development Strategies, Poverty Reduction Strategies, Environmental and Social Impact Assessments, the Kyoto Protocol or the Convention on Biological Biodiversity. Coordination therefore is required among different government bodies (e.g. Ministry of Agriculture, Energy, Environment, Industry, Trade, etc.), levels and actors.

Finally, at the international level, cooperation is also crucial for the development of a sugarcane bioethanol industry oriented towards poverty reduction and environmental sustainability. South-South cooperation can play an important role in overcoming many of the technical challenges. Countries can benefit from the technical and scientific knowledge of Brazil, which is at the forefront of the industry. One example in that sense is the illustrated by the Brazil-UK-Africa Partnership for bioethanol development. International financial institutions can help, for example, by mitigating political risk for project development in developing countries. Elimination of trade barriers is another issue to be addressed by governments to enhance development opportunities associated with sugarcane bioethanol. This would be also aligned with the last Millennium Development Goal that calls to 'develop a global partnership for development'.

References

Bakker, S.J.A., 2006. CDM and biofuels - Can the CDM assist biofuels production and deployment? Energy Research Centre of the Netherlands, October. Available at: http://www.senternovem.nl/mmfiles/CDMandBiofuels_tcm24-222843.pdf

Bolling, C. and Suarez, N., 2001. The Brazilian sugar industry: recent developments. In: Sugar and Sweetener Situation & Outlook /SSS-232/September.

Cartwright, A., 2007. Biofuels trade and sustainable development: An analysis of South African bioethanol. Working document, International Institute for Environment and Development, London, UK.

Cloin, J., 2007. Liquid biofuels in Pacific island countries. SOPAC Miscellaneous Report 628. Pacific Islands Applied Geoscience Commission, Suva, Fiji.

Coelho, S.T, 2005. Biofuels – Advantages and Trade Barriers. Prepared for the United Nations Conference on Trade and Development, February, Geneva, Switzerland.

Dufey, A., 2006. Biofuels production, trade and sustainable development: emerging issues. Sustainable Markets Discussion Paper 2. International Institute for Environment and Development, London, UK. Available at: http://www.iied.org/pubs/pdf/full/15504IIED.pdf

Dufey, A., 2007. International trade in biofuels: Good for development? and Good for the Environment? IIED Policy Briefing. International Institute for Environment and Development, London, UK. Available at: http://www.iied.org/pubs/pdf/full/11068IIED.pdf

Dufey, A., M. Ferreira and L. Togeiro, 2007a. Capacity Building in Trade and Environment in the Sugar/Bioethanol Sector in Brazil. UK Department for Environment, Food and Rural Affairs, London, UK.

Dufey, A., S. Vermeulen and W. Vorley, 2007b. Biofuels: Strategic Choices for Commodity Dependent Developing Countries. Common Fund for Commodities, September.

Echeverri-Campuzano, H., 2000. Columbia Paving the Way in Renewable Fuels for Transport in Renewable Energy for Development. Stockholm Environment Institute –Newsletter of the Sustainable Energy Programme, October 2000 Vol. 13 No. 3.

Echeverri-Campuzano, H., 2002. Fuel Ethanol Program in Colombia, CORPODIB. Available at: http://www.iea.org/textbase/work/2002/ccv/ccv1%20echeverri.pdf

ESMAP, 2005a. The Vulnerability of African Countries to Oil Price Shocks: Major Factors and Policy Options - The Case of Oil Importing Countries Energy Sector Management Assistance Programme, Report 308/05. The International Bank for Reconstruction and Development, Washington DC, August. Available at: http://wbln0018.worldbank.org/esmap/site.nsf/files/308-05+Final_to_Printshop.pdf/$FILE/308-05+Final_to_Printshop.pdf

ESMAP, 2005b. Potential for biofuels for transport in developing countries. Energy Sector Management Assistance Programme, World Bank, Washington DC, USA. Available at: http://esmap.org/filez/pubs/31205BiofuelsforWeb.pdf

ESMAP, 2006. Annual Report 2005. Energy Sector Management Assistance Programme, The International Bank for Reconstruction and Development, Washington DC, June. Available at: http://www.esmap.org/filez/pubs/ESMAP05.pdf

E4tech, 2006. Feasibility study on the production of bioethanol from sugarcane in the Caribbean Region. Final report for DFID Caribbean, DFID, March.

FAO, 2007. A Review of the Current State of Bioenergy Development in G8 +5 Countries. Global Bioenergy Partnership, Food and Agriculture Organization of the United Nations, Rome.

Gonsalves, J. 2006a. An Assessment of the Biofuels Industry in India. United Nations Conference on Trade and Development, UNCTAD/DITC/TED/2006/6.

Gonsalves, J. 2006b. An Assessment of the Biofuels Industry in Thailand. United Nations Conference on Trade and Development, UNCTAD/DITC/TED/2006/7.

Hazell, P., G. Shields and D. Shields, 2005. The nature and extent of domestic sources of food price instability and risk. Paper presented to the workshop 'Managing Food Price Risk Instability in Low-Income Countries' February 28 – March 1, 2005, Washington DC, USA.

Horta, L., 2006. Potencial Económico y Ambiental del Etanol como Oxigenante en la Gasolina. Convenio Costarricense-Alemán de Cooperación Técnica, Proyecto Aire Limpio San José. MOPT, MS, MINAE y GTZ. Costa Rica.

Henniges, O. and J. Zeddies, 2006. Bioenergy in Europe: Experiences and Prospects in Bioenergy and Agriculture: Promises and Challenges, IFPRI, FOCUS 14, Brief 9 of 12, December. Available at: http://ifpri.org/2020/focus/focus14/focus14_09.pdf

ICRISAT, 2007. Pro-poor biofuels outlook for Asia and Africa: ICRISAT's perspective. International Crops Research Institute for the Semi-Arid Tropics, Nairobi, Kenya.

IEA, 2004. Biofuels for Transport: An International Perspective. International Energy Agency, Paris, April.

IFPRI, 2008. What goes down must come up: Global Food Prices Reach New Heights. IFPRI Forum, International Food Policy Research Institute, March. Available at: http://www.ifpri.org/PUBS/newsletters/IFPRIForum/if21.pdf

Jackson, M., 2004. Ethanol Fuel Use - Promising Prospects for the Future in Renewable Energy for Development. Stockholm Environment Institute –Newsletter of the Sustainable Energy Programme, January 2004 Vol. 17 No. 1.

Johnson, F. and F. Rosillo-Calle, 2007. Biomass, Livelihoods and International Trade - Challenges and Opportunities for the EU and Southern Africa. Stockholm Environment Institute, Climate and Energy Report 2007-01.

Johnson, F., V. Seebaluck, H. Watson and J. Woods, 2006. Bioethanol from sugarcane and sweet sorghum in Southern Africa: Agro-Industrial Development, Import Substitution and Export Diversification in Linking Trade, Climate Change and Energy. ICTSD Trade and Sustainable Energy Series, International Centre for Trade and Sustainable Development, Geneva, Switzerland, November.

Kartha, S., 2006. Environmental effects of bioenergy. In: Hazell, P. and Pachauri, R.K. (eds) Bioenergy and agriculture: promises and challenges. 2020 Vision Focus 14. International Food Policy Research Institute, Washington DC, USA, and The Energy and Resources Institute, New Delhi, India.

Koplow, D., 2006. Biofuels: At What Cost? Government Support for Ethanol and Biodiesel. In: the United States report prepared by Earth Track, Inc for The Global Subsidies Initiative (GSI) of the International Institute for Sustainable Development (IISD), Geneva, Switzerland.

Kutas, G. and C. Lindberg, 2007, Biofuels – At what Cost? Government Support for Ethanol and Biodiesel in the European Union for the Global Subsidies Initiative (GSI). IISD, Geneva, Switzerland.

Lambe, F., 2008. Development of Local Biofuel Markets: A Case Study from Ethiopia. In: Renewable Energy for Development. Stockholm Environment Institute –Newsletter of the Sustainable Energy Programme, May 2008, Vol. 21 No. 1.

Macedo I.C (Ed.). 2005. A Energia da Cana de Açúcar - Doze estudos sobre a agroindústria da cana-de-açúcar no Brasil e a sua sustentabilidade UNICA.

Mayers, J. and S. Vermeulen, 2002. From raw deals to mutual gains: company-community partnerships in forestry. Instruments for Sustainable Private Sector Forestry series. International Institute for Environment and Development, London, UK.

Ministerio da Agricultura, Pecuaria e Abastecimiento and Assessoria de Gestao Estrategica, 2006. Agribusiness Projections: World and Brazil, Executive Summary available at: http://www.agricultura.gov.br/pls/portal/docs/PAGE/MAPA/MENU_LATERAL/AGRICULTURA_PECUARIA/PROJECOES_AGRONEGOCIO/COPY_OF_PROJECOES_AGRONEGOCIO/EXECUTIVE%20SUMMARY%20AGRIBUSINESS%20PROJECTIONS_0.PDF

Murillo, C., 2007. Biofuels trade and sustainable development: the case of Costa Rica. Working document, International Institute for Environment and Development, London, UK.

Murphy, S., 2007. The multilateral trade and investment context for biofuels: Issues and challenges. International Institute for Environment and Development, London/Institute for Agriculture and Trade Policy, Minneapolis.

OECD-FAO, 2007. OECD & FAO Agricultural Outlook 2007-2016. Organisation for Economic Cooperation and Development, Paris, France, and Food and Agriculture Organisation of the United Nations, Rome, Italy.

OXFAM, 2008. Another Inconvenient Truth – how biofuels policies are deepening poverty and accelerating climate change. Oxfam Briefing Paper, Oxfam International, June.

Peskett, L., R. Slater, C. Stevens and A. Dufey, 2007. Biofuels, agriculture and poverty reduction. Report prepared for the UK Department for International Development. Overseas Development Institute, London, UK.

Rafi Khan, S., M. Yusuf, S. Adam Khan and R. Abbasy, 2007. Biofuels Trade and Sustainable Development: The Case of Pakistan. Working Document International Institute for Environment and Development, London, UK.

Rosegrant, M., 2008. Biofuels and Grain Prices: Impacts and Policy Responses, International Food Policy Research Institute May. Available at: http://www.ifpri.org/pubs/testimony/rosegrant20080507.pdf

Rossi, A. and Y. Lambrou, 2008. Gender and Equity Issues in Liquid Biofuels Production - Minimizing the Risks to Maximize the Opportunities. Food and Agriculture Organization of the United Nations, Rome. Available at: ftp://ftp.fao.org/docrep/fao/010/ai503e/ai503e00.pdf

Schlamadinger, B. and I. Jürgens, 2004. Bioenergy and the Clean Development Mechanism. Joint International Energy Association and FAO policy paper, IEA Bioenergy Task 38, Paris, International Energy Agency. Available at http://www.ieabioenergy-task38.org/publications/.

Schmidhuber, J., 2007. Biofuels: an emerging threat to Europe's food security? Notre Europe Policy Paper 27. Notre Europe, Paris, France.

Severinghaus, J., 2005. Why we import Brazilian bioethanol. Iowa Farm Bureau.

Smeets, E., M. Junginger, A. Faaij, A. Walter and P. Dolzan, 2006. Sustainability of Brazilian bio-ethanol. Copernicus Institute - Department of Science, Technology and Society for SenterNovem, The Netherlands Agency for Sustainable Development and Innovation, The Netherlands, Utrecht.

Stern, N., 2006 The Economics of Climate change: Stern Review. Cambridge University Press, Cambridge.

UN-Energy, 2007. Sustainable bioenergy: a framework for decision makers. United Nations. Available at: http://esa.un.org/un-energy/pdf/susdev.Biofuels.FAO.pdf

Woods, J. and P. Read, 2005. Arguments for Bioenergy Development in Policy Debate on Global Biofuels Development. Renewable Energy Partnerships for Poverty Eradication and Sustainable Development, June 2005, Partners for Africa/Stockholm Environment Institute.

World Bank, 2008. Rising Food Prices: Policy Options and World Bank Response. World Bank.

Worldwatch Institute, 2006. Biofuels for transportation: global potential and implications for sustainable agriculture and energy in the 21st century. Report prepared for the German Federal Ministry of Food, Agriculture and Consumer Protection (BMELV). Worldwatch Institute, Washington DC, USA.

Zarrilli, S., 2006 Trade and Sustainable Development Implications of the Emerging Biofuels Marke'. In: International Centre for Trade and Sustainable Development Linking Trade, Climate Change and Energy: Selected Issue Briefs. International Centre for Trade and Sustainable Development, Geneva, Switzerland.

Chapter 10
Why are current food prices so high?

Martin Banse, Peter Nowicki and Hans van Meijl

1. World agricultural prices in a historical perspective

World agricultural prices are very volatile which is due to traditional characteristics of agricultural markets such as inelastic (short run) supply and demand curves (see, Meijl *et al.* 2003).[13] The volatility is also high because the world market is a relatively small residual market in a world distorted by agricultural policies.[14] The combination of high technological change and inelastic demand cause real world prices to decline in the long run (trend). The prices, however, of many (major) agricultural commodities have risen quickly over recent years (see Figure 1).

Recent increase in agricultural prices are strong, but even with the increase that we have observed in the last three years, real agricultural prices are still low compared to the peaks in prices of the mid-70s. Local prices are linked with these world prices. The transmission effect depends on the transparency of markets, market power and accessibility

[13] 'World food prices are instable and will remain unstable in the future. Forecast errors are large in predictions of world prices. There are always unexpected events in important drivers such as yields which are dependent on weather, plagues and diseases' (See Van Meijl *et al.*, 2003).

[14] Trade share (2006) in global production: rice (7%), cheese (7%), coarse grains (11%) and wheat (20%), FAO Statistics.

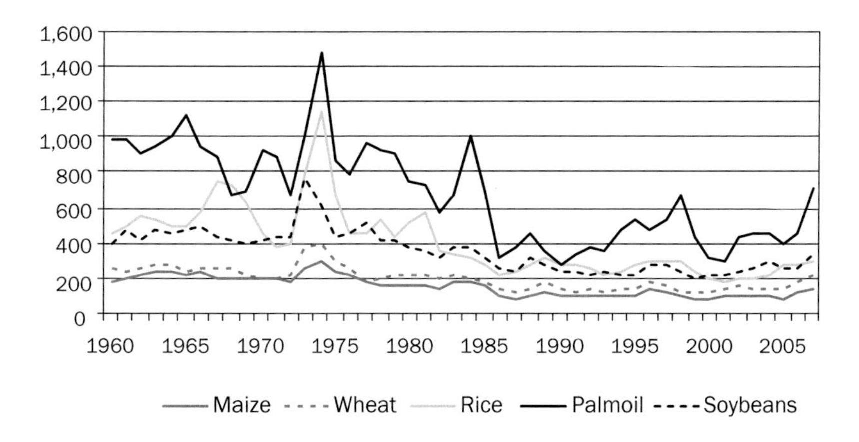

Figure 1. Development of world agricultural prices, 1960–2007, USD/ton, in constant USD (1990). Source: World Bank data base (2008).

Figure 2 depicts the price index for food commodities along with an index for the average of all commodities and an index for crude oil. Although the food commodity index has risen more than 60 percent in the last 2 years, the index for all commodities has also risen 60 percent and the index for crude oil has risen even more (see also Trostle, 2008). Since 1999 food commodity prices have risen 98 percent (as of March 2008); the index for all commodities has risen 286 percent; and the index for crude oil has risen 547 percent. In this perspective, the recent rise in food commodity prices is moderate. Figure 3 shows that spot prices in early 2008 for soybean and wheat are declining again while the spot prices for rice and crude oil continue to rise. The prices of wheat and soybeans declined by almost 30% and almost 20%, respectively, since their peak at the end of February this year.

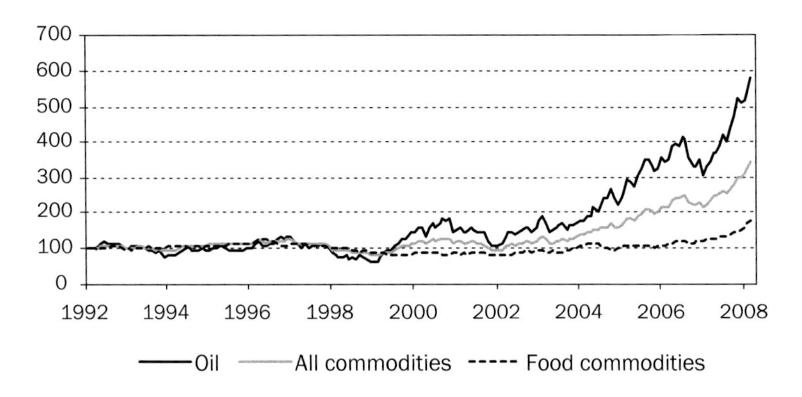

Figure 2. Index of oil, food and all commodities, 1992-2008, January 1992=100. Source: International Monetary Fund: International Financial Statistics.

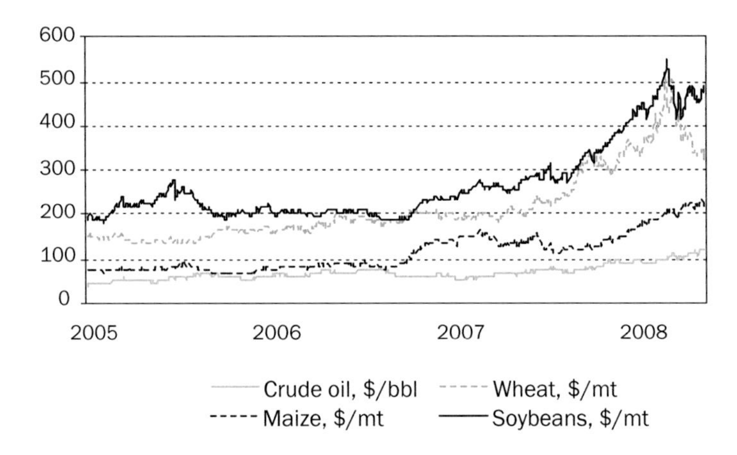

Figure 3. Daily price notations for crude oil, wheat, maize and soybeans; spot prices, 2005-2008, at current USD. Source: World Bank data base (2008) from January, 1 2005 to May, 15 2008.

However, although real food prices are not extremely high in a historical perspective and other commodities have risen more, an increase in the price of food – a basic necessity – causes hardships for many lower income consumers around the world. This makes food-price inflation socially and politically sensitive. This is why much of the world's attention is now focused on the increase in food prices more than on the more rapid increase in prices of other commodities, (see Trostle, 2008: 4).

The question on the minds of many consumers around the world is, 'Will food prices drop again this time?' Or, stated another way, 'Is the current price spike any different from those of the past, and if so, why?'

2. Long run effects

2.1. Long run drivers of demand[15]

Population and macro-economic growth are important drivers of demand for agricultural products. In past years, rapid population growth has accounted for the bulk of the increase in food demand for agricultural products, with a smaller effect from income changes and other factors (Nowicki *et al.*, 2006)[16]. The world's population growth will fall to about 1% in the coming ten years. Continued economic growth is expected over the coming period in almost all regions of the world and this driver of demand will become more important than population growth in the future (see Figure 4).

2.2. Expected population developments in period 2005-2020

- The world's population growth will fall from 1.4% in the 1990-2003 period to about 1% in the coming ten years. This is mainly due to birth or fertility rates, which are declining and are expected to continue to do so.
- Almost all annual population growth will occur in low and middle income countries, whose population growth rates are much higher than those in high income countries.
- Europe's share in world population has declined sharply and is projected to continue declining during the 21st century.
- Population growth in Europe is very low (0.3% yearly for EU-15: old EU member states) or slightly negative (-0.2% for EU-10: new EU member states).
- The uncertainty with regard to birth and death rates at world or regional level is not too large. However, migration flows between countries and regions are much more uncertain.

[15] Based on Scenar 2020 (Nowicki *et al.*, 2006).

[16] Projections for population and GDP for the EU member states are taken from a study of the Economic Policy Committee of the European Commission called 'The 2005 EPC projection of age-related expenditure: agreed underlying assumptions and projections methodologies, 2005'. The projections for the rest of the world are based on assumptions used in the OECD and USDA agricultural Outlooks.

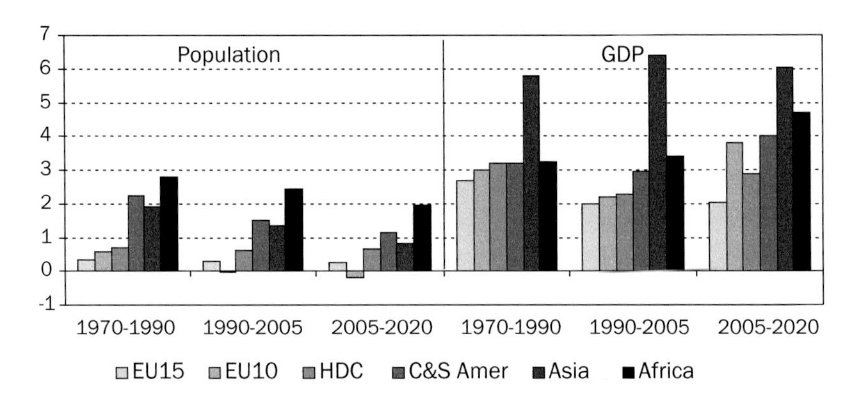

Figure 4. World population and GDP growth (annual growth %). Source: USDA for 1970-1990 and 1990-2005. Projections for 2005-2020 derived from Scenar 2020, Nowicki *et al.* (2006). HDC = High Income Developed Countries, C&S Amer = Central and South America

2.3. Global income growth

- Robust economic growth is expected over the coming period in almost all regions of the world in the baseline scenario (see Figure 4).
- Economic growth will be considerably higher for most of the transitional and developing countries than for the EU-15, the United States and Japan, in particular for Brazil, China, India and the new EU member states. Incomes in Europe are expected to increase slightly over the coming years.
- Annual income growth in Europe is about 2% for EU-15 and 3.8% for EU-10.
- World and EU economic growth in the future stays uncertain and depends on the amount of investments in education and research, on technological opportunities, on the degree of (labour) participation in the political, societal and market arenas, and on the liberalisation of world commodity and factor markets.

The robust growth of income per capita leads to more 'luxury' consumption in developed countries. This implies more convenience food, processed products (ready to eat) and food safety, environmental and health concerns. In developed countries the total amount of food consumed will only grow in a limited manner. However, in developing countries a higher income induces more consumption and a shift to more value-added products. Important is the switch from cereals to meat consumption, as an increased demand for meat induces a relatively higher demand for grain and protein feed. To produce 1 kg of chicken, pork and beef, respectively 2.5 kg, 6.5 kg and 7 kg of feed are required.[17]

[17] The numbers describe upper-bound estimates of conversion rates: 7 kg of maize to produce 1 kg of beef, 6.5 kg of maize to produce 1 kg of pork, and 2.6 kg of maize to produce 1 kg of chicken (Leibtag, 2008). Modern technology, however, require much less feed especially in pork production; here average feed conversion rates are between 3.2-2.6 kg of feed per kg of meat.

2.4 Long-term drivers of supply

With regard to grain and oilseed production, yield and area developments are important drivers of supply. Figure 5 shows that production growth was almost totally determined by yield increase while the total area harvested was more or less constant. The growth in yields declined from 2% per year in the 1970-1990 period to 1.1% in the 1990-2007 period. USDA expects the growth to decline to 0.8% per year for the period 2009-2017 (USDA, 2008). At the global scale, crop production area increased in the 1970-2007 period by 0.15% per year, and USDA expects the area to grow by 0.4% per year in the period 2007-2017.

Figure 6 shows that growth rates of yields for major cereals in developing countries are slowing. It should be mentioned that the decline in annual growth rates is not necessarily related to a decline in absolute yield growth per annum. An important explanation for the decreasing yield growth rates might be the declining public agricultural research and development spending over time in both developing and developed countries (Figure 7). Although private sector research has grown, private sector R&D is mostly cost reducing\ short run oriented instead of public R&D, which is often more yield enhancing\long term oriented.

- The direct link between R&D spending and yield growth had been intensively discussed amongst agricultural scientists and is not fully clear.
- The general outcome of this discussion is that an additional growth in yield rates requires more than additional spending in capital stock but also investment in human capital stock and improvements in market institutions

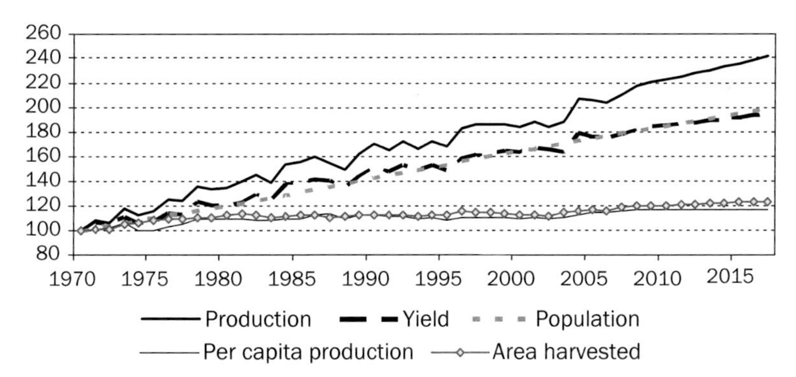

Figure 5. Development of world grain and oilseed production. Source: USDA Agricultural Projections to 2017.

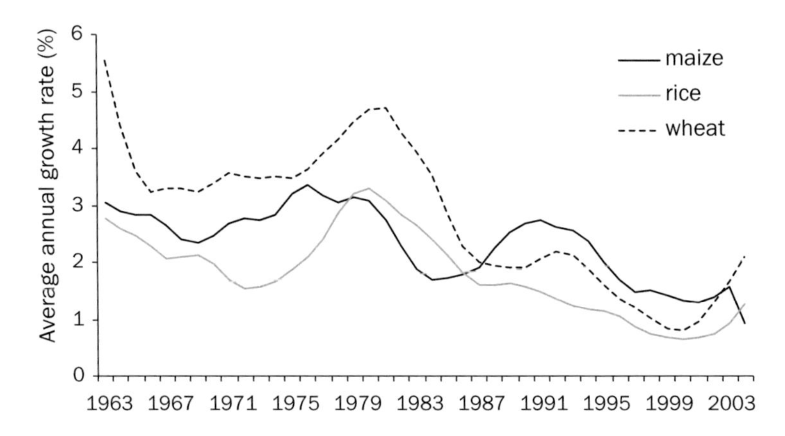

Figure 6. Development annual yields for selected cereals in developing countries. Source: World Development Report 2008.

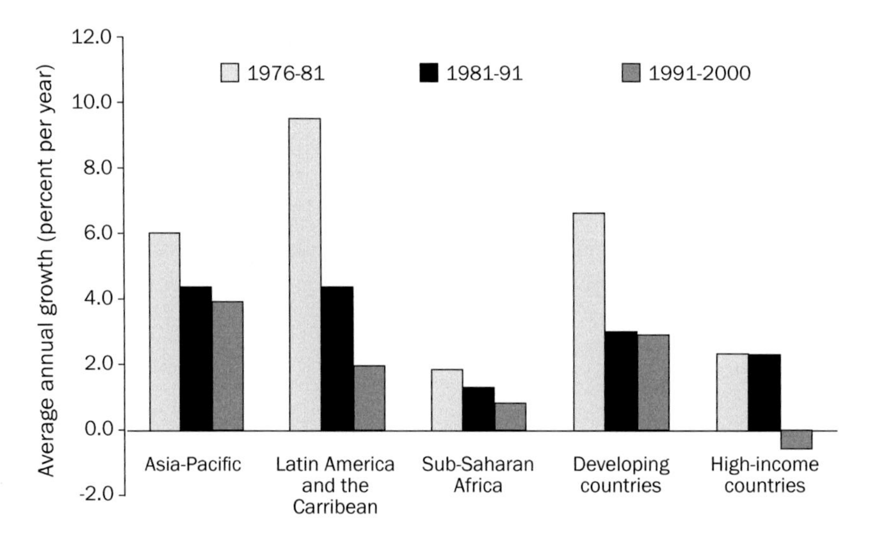

Figure 7. Public Agricultural R&D Spending Trends, 1976-2000. Source: Pardey *et al.* (2006).

3. What explains the recent increase in agricultural prices?

A combination of record low global inventory levels, weather induced supply side shocks, surging outside investor influence, record oil prices and structural changes in demand for grains and oilseeds due to biofuels have created the high prices. The question is whether it is a coincidence that the past and current high price levels coincide with high oil prices or whether other reasons for the current price peak are more important.

3.1. Effects on the supply side

As mentioned above the variation of yields due to climatic conditions, the development of input prices – fertilizer, diesel and pesticides – as well as the level of political support are the main drivers of supply. The following items provide some information on these points (Figure 8):

- Poor harvests in Australia, Ukraine and Europe for wheat and barley. According to FAO statistics, these three regions contributed on average 51% of total world barley production and 27% of total world wheat production for the period 2005-2006.
- Lower harvests in wheat and barley are more than compensated by a bumper harvest for maize worldwide.
  - Therefore, world cereal production increased in total even in 2007.
  - The bumper harvest in maize kept maize prices low and the wheat-maize spread increased significantly (Figure 3).
  - Only recently have maize prices also strongly increased.
- Higher energy prices lead to higher food prices as costs (e.g. fertilizer, processing, and transport) increase. Higher transport costs induce higher price effects as distances increase.
- CAP policies such as mandatory set-aside regulation or production quota restrained supply. Furthermore, there was a change from price to income support and compensatory payments became decoupled, set aside was introduced and export subsidies were diminished. Some of these measures limited supply within the EU. However, the general aim of the last CAP reforms was an enforcement of farmers' ability to react to market signals instead of following policy signals given by market price support. Measures

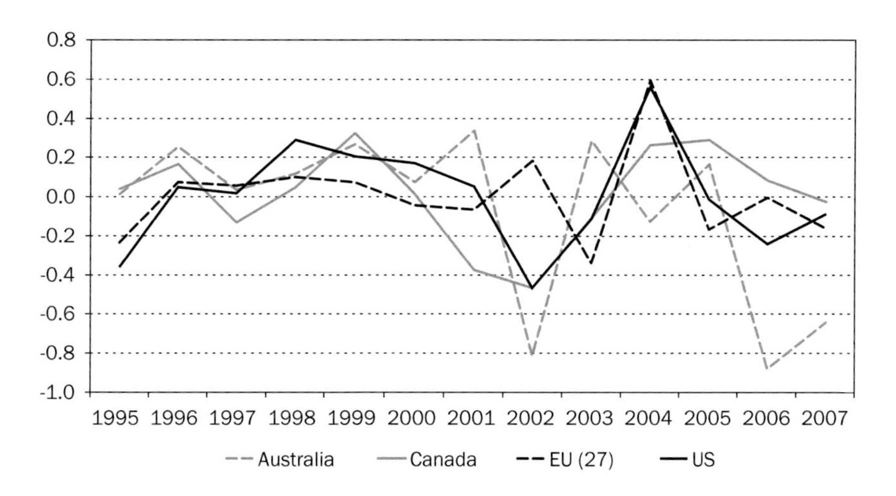

Figure 8. Deviation from trend in yields (wheat and coarse grains) in tons/ha. Source: OECD-FAO Agricultural Outlook 2008-2017 (2008).

aimed to restrict supply, e.g. production quota or set-aside requirements, are instruments designed for a world with declining prices, but which may act to reinforce prices in case of food shortages.
- Low prices in the last decades did not provide an incentive to invest in productivity enhancing technologies.

3.2. Effects on the demand side

Compared to the variability of agricultural supply, the demand of agri-food products is rather inelastic. For most agricultural commodities price and income elasticities are small, i.e. long-term demand for primary agricultural products is more determined by population growth and less by income growth. Within the last years the demand for agri-food products have been determined by the following driver:
- Constant demand in Europe and Northern America with an increase in demand in Asian countries
- Change in diet in emerging economies.
- Additional demand for biofuels:
  - 5% of global oilseed production is processed to biodiesel or is used directly for transportation.
  - 4.5% of global cereal production is used for ethanol production.
  - Therefore, this marginal extra demand triggered the markets.
  - However, biofuels are not new. Ethanol based on sugarcane exists in an economically profitable way in Brazil for a long time.
  - Increasing food and feedstock prices make biofuels less profitable and food more profitable. This shifts production back to food (in US is this already visible; Trostle, 2008, p.17). With current high prices for soybeans in the US margins for biodiesel became already negative and the biodiesel production slowed down [see presentation of Gerald A. Bange (USDA) on the Agricultural Markets Roundtable held April 22, 2008 Washington, DC at the Commodity Futures Trading Commission].

The development of both – supply and demand side – contribute to the development of stocks which is illustrated in the following Figure 9. The trend of a declining stock to use ratio as has increased and stocks for wheat are currently running on empty. This implies that all the shocks mentioned above could not be mitigated by using stocks but lead immediately to price increases. Furthermore, it enabled speculation (with stocks available there would have been less room for speculation)

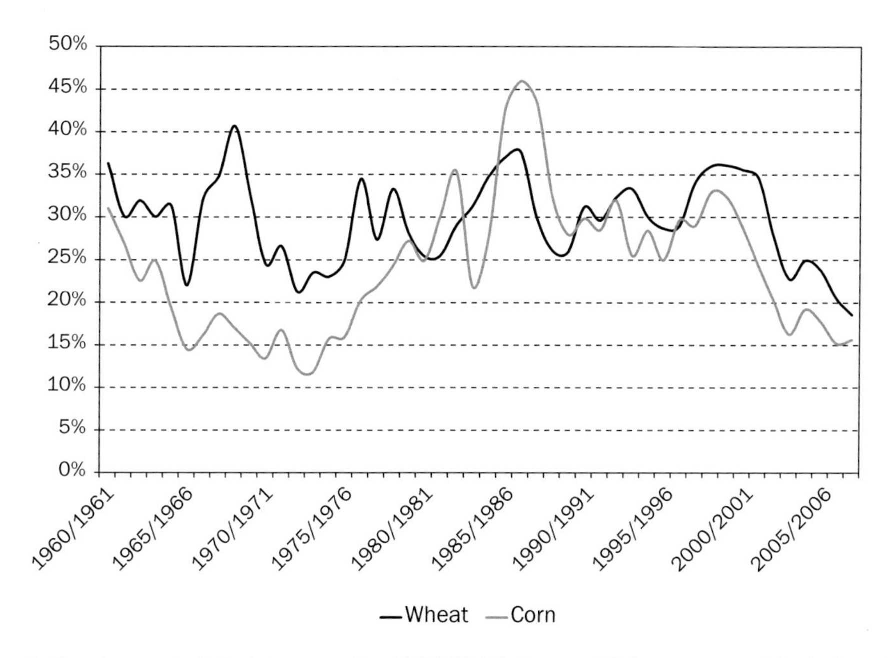

Figure 9. Development of stock to use ratio, 1960-2007. Source: US Department of Agriculture PSD View database, June 2008.

3.3 Policy responses to rising food prices

- The rapidly increasing world prices for food grains, feed grains, oilseeds, and vegetable oils are causing domestic food prices at the consumer level to rise in many countries. In response to rising food prices, some countries are beginning to take protective policy measures designed to reduce the impact of rising world food commodity prices on their own consumers. However, such measures typically force greater adjustments and higher prices onto global markets.
- In the fall of 2007, some exporting countries made policy changes designed to discourage exports so as to keep domestic production within the country. The objective was to increase domestic food supplies and restrain increases in food prices. Table 1 depicts a partial list of these policy changes.

Table 1. Policy responses to rising food prices.

Eliminated export subsidies:

China eliminated rebates on value-added taxes on exported grains and grain products. The rebate was effectively an export subsidy that was eliminated.

Export taxes:

China, with food prices still rising after eliminating the value-added tax rebate, imposed an export tax on a similar list of grains and products.

Argentina raised export taxes on wheat, maize, soybeans, soybean meal, and soybean oil.

Russia and Kazakhstan raised export taxes on wheat.

Malaysia imposed export taxes on palm oil.

Export quantitative restrictions:

Argentina restricted the volume of wheat that could be exported even before raising export taxes on grains.

Ukraine established quantitative restrictions on wheat exports.

India and Vietnam put quantitative restrictions on rice exports.

Export bans:

Ukraine, Serbia, and India banned wheat exports.

Egypt, Cambodia, Vietnam, and Indonesia banned rice exports. India, the world's third largest rice exporter, banned exports of rice other than basmati, significantly reducing global exportable supplies.

Kazakhstan banned exports of oilseeds and vegetable oils. Early in 2008, importing countries also began to take protective policy measures to combat rising food prices. Their objective was to make high-cost imports available to consumers at lower prices. A partial list of policy changes follows.

The following countries reduced import tariffs:

India (wheat flour).

Indonesia (soybeans and wheat; streamlined the process for importing wheat flour).

Serbia (wheat).

Thailand (pork).

EU (grains).

Korea and Mongolia (various food commodities)

Subsidizing consumers:

Some countries, including Morocco and Venezuela, buy food commodities at high world prices and subsidize their distribution to consumers.

Other decisions by importers:

Iran imported maize from the United States, something that has occurred rarely – only when they could not procure maize elsewhere at reasonable prices.

The policies adopted by importing countries also changed price relationships in world markets. Their policy changes increased the global demand for food commodities even when world prices were already rapidly escalating.

Source: Trostle (2008).

3.4. Other effects

- USD exchange rate developments. World prices are denominated in dollars and the dollar depreciated against most currencies. The increase in prices in other currencies is therefore much less.

Speculation:

- In recent months spot and future prices do not fully converge.
- Future prices remain higher than prices on spot markets.
  - Reason for this development:
    - Most hedging (90%) is Index-hedging, i.e. 'traditional' short- and long hedging does not dominate the price development in the future markets.
    - Thus, if everybody expects high prices, then future prices tend to be higher than the spot prices.
  - So, part of current high prices can be attributed to this 'bubble'.
- Difficult to estimate the impact of speculation in this story.
  - The crises on the financial markets are diverting funds away from traditional financial institutions leading to a large pool of funds available for investments in other markets.
  - There is definitely a impact of speculation in current high prices
  - Hard to say it makes X %.
  - Growing volatility in food markets due to the fact that most of hedging is based on index funds and not anymore on the 'traditional' short and long hedging. This share is less than 10% in total market volume.
  - An example for the current volatility: In the 1st week of March the fluctuation of maize prices was more than 150 USD/t, which is more than last year's average maize price!
- Impact of speculation on current spike in agricultural prices is difficult to quantify. Figure 10 shows the composition of the maize futures markets broken down between commercial merchants, managed money funds and commodity index traders together with the price development in USD per bushel of maize (right-hand scale).
  - It clearly shows that not only the 'speculative' index and fund hedging but also the increase in short futures by commercial merchants contributed to the dramatic increase in maize future prices.
  - However, the managed money funds which are mostly pension funds – which diversify their portfolio now also to agricultural commodities – cut down their purchase of additional contracts on long position when prices increased dramatically (Figure 10).
  - A formal assessment is hampered by data and methodological problems, including the difficulty of identifying speculative and hedging-related trades.

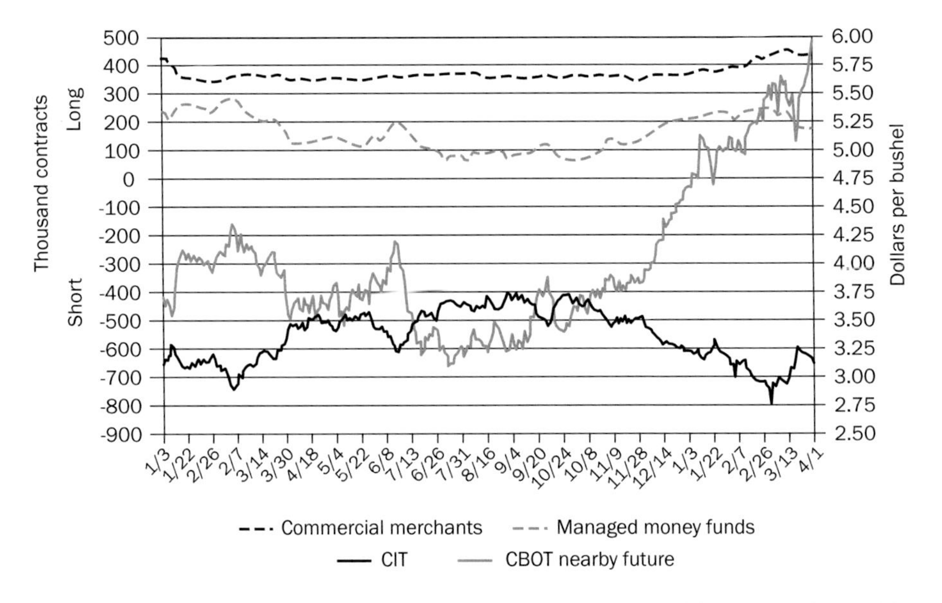

Figure 10. CBOT Corn Market Composition January 2007 – April 2008. Source: Derived from a presentation of Dave Kass at the Agricultural Markets Roundtable held April 22, 2008 Washington, DC at the Commodity Futures Trading Commission.

- A number of recent studies seem to suggest that speculation has not systematically contributed to higher commodity prices or increased price volatility.
  - For example, a recent IMF staff analysis (September 2006 World Economic Outlook) shows that speculative activity tends to respond to price movements (rather than the other way around), suggesting that the causality runs from prices to changes in speculative positions.
  - The Commodity Futures Trading Commission has argued that speculation may have reduced price volatility by increasing market liquidity, which allowed market participants to adjust their portfolios, thereby encouraging entry by new participants.

4. First quantitative results of the analysis of key driving factors

- OECD Outlook 2007-2017: The OECD performed some scenarios to see the impact of various drivers on their Outlook projection (OECD-FAO, 2008). This analysis highlights the outcome of a situation where biofuel policies are in place under the reference scenario and different assumptions are moderate, e.g. income growth, development of crude oil prices, etc.:
  - If biofuel production stays at its 2007 level, then world wheat prices would be 5% lower, maize 13% lower and vegetable oil 15% lower compared to the reference scenario where biofuel production in 2017 more than doubles relative to the 2007 level.

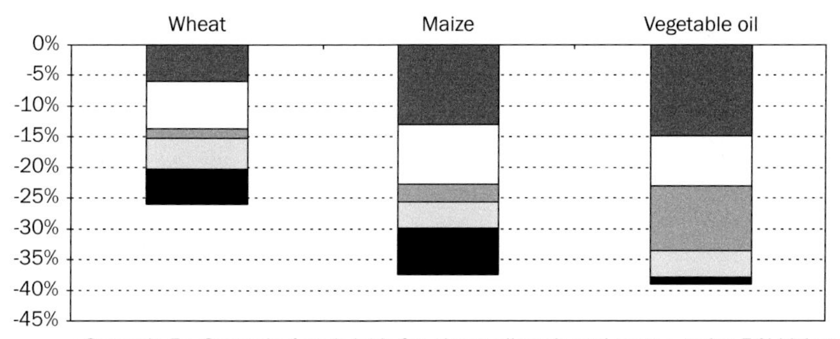

Figure 11. Sensitivity on analysis of world price changes. Source: OECD-FAO Agricultural Outlook 2008-2017. Highlights. (2008).

- A constant crude oil price implies 10% lower prices for all three commodities, due to the fact that the assumed high crude oil price under the reference scenario will make biofuel crops more profitable.
- Lower income growth is especially relevant for vegetable oils (more than 10%).
- A stronger US dollar of 10% leads to about 5% lower prices for wheat, maize and vegetable oil relative to the baseline.
- Higher growth rates in yields for important biofuel crops will lower the world market prices for their production by more than 5% for wheat and maize.

These results are inline with our own results on the impact of biofuel policies, which are presented in Figure 12.

- International Food Policy Research Institute (IFPRI) study (e.g. Von Braun *et al.*, 2008).
  - The percentage contribution of biofuels demand to price increases from 2000-07 is the difference between 2007 prices in the two scenarios, divided by the increase in prices in the baseline from 2000-2007.
  - The increased biofuel demand between 2000 and 2007, compared with previous historical rates of growth, is estimated to have accounted for 30 percent of the increase in weighted average cereal prices during 2000-07.
    - › Maize – 39%.
    - › Rice – 21%.
    - › Wheat – 22%.

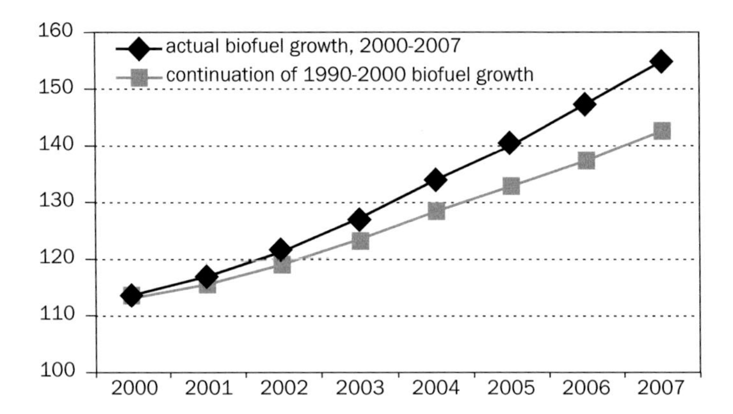

Figure 12. Biofuels: Impact on world cereal prices since 2000. Source: Impact Simulations 2008. IFPRI.

- Rapid growth in biofuel demand has contributed to the rapid rise in cereal prices, but it has not been a dominant driving force in the 2000-07 period, except perhaps in the case of maize.
- The fundamentals of supply and demand seem to be playing more of a role in the rapid increase in prices during this period, especially for commodities like rice and wheat.
- After 2007 prices increases – for rice in particular – seem to be driven by the relatively 'thin' nature of the rice market with a limited amount of international trade compared to total production.
- Unilateral trade policy actions of individual Asian countries, which have sought to put into place export bans and import subsidies for rice.
- Speculative trading and storage behaviour; private operators taking advantage of opportunities.

- Agri-Canada quantified the impact of all the policy responses (Figure 13). The impact of policies added a few percent for almost all commodities, except for rice where the impact is substantial (16%).

Experts are pointing out that it is hard to quantify the separate impacts. The contribution of biofuel demand to the increase in average cereal prices of 30% presented by IFPRI was criticized by some colleagues. Some find it too high, other too low. However, all studies point out that a combination of factors was responsible for the rise. The analyses of OECD, FAPRI and also of Banse *et al.* (2008a,b) indicated that the impact on world price levels is commodity specific. For maize the impact is relatively high due to the fact that most US ethanol production is maize-based. For other cereals – e.g. wheat and rice, where the use for biofuels is almost zero – only indirect effects over the land use affects the world price level. For those commodities an estimated increase of 30% – as indicated in the IFPRI estimates – seems to be rather high.

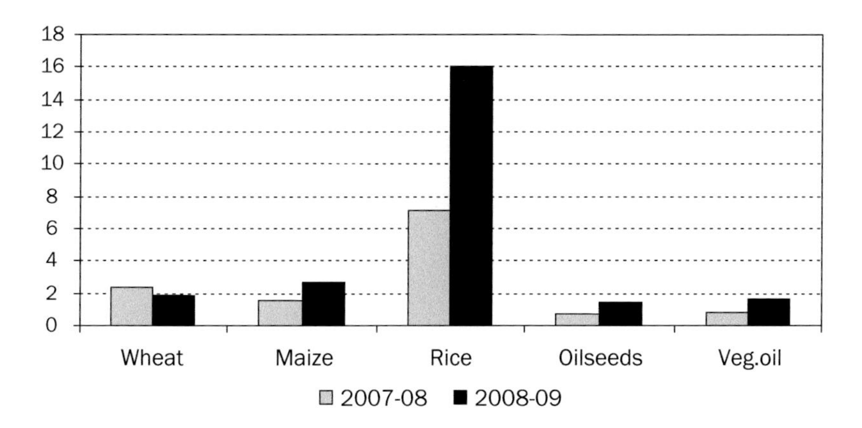

Figure 13. Impact of export restriction policies on world prices. Source: Agriculture and Agri-Food Canada, unpublished.

5. The future

After the discussion of those driving elements which contributed to the current spike in food prices this section depicts some elements which might contribute to the long-term development of agri-food prices. This sections also allows to identify possible solutions for the current crisis on world food markets.

- High prices are their own worst enemy. Increased profit margins entice entrepreneurial investment, which results in increased production. Lower market prices inevitably follow. The 'invisible hand' of Adam Smith ensures that winners' gains and losers' losses will be temporary, as entrepreneurs correct market imbalances. In the USA, in the 2008 spring planting farmers are shifting from maize to wheat and soybeans, setting the prices of the latter on a downward trajectory and stabilising the price of the former.
- Higher prices induce more production as planted areas increase and available arable land will be used more intensively. Therefore, the current situation is not structural and as a result prices will go down again. However, first stocks have to be built up again. Both effects take some time. In Brazil and Russia there are ample opportunities as additional land can be taken into production, whereas in many other countries production can only be higher due to intensification. According to USDA analyses, Russia, Ukraine and Argentina can become one of the world's top grain exporters.
- R&D investments in agriculture (e.g. yields, etc.) become more profitable with higher food prices.
- Strategic stocks are essential to limit price volatility in world agricultural markets, but they are costly.
- The expected impact on world prices of the 10% EU-biofuel directive and the various global biofuel initiatives is depicted in the graph below (Banse *et al.*, 2008a,b). If all initiatives are implemented together and technological change stays on the historic trend,

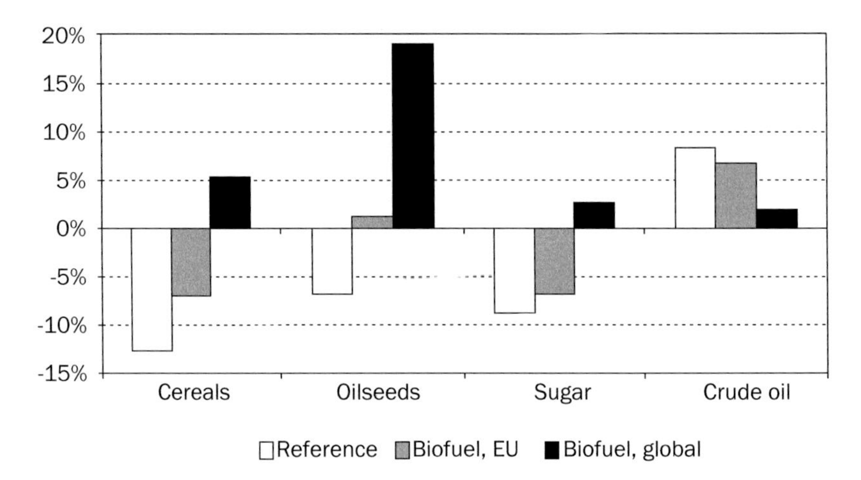

Figure 14. Change in real world prices, in percent, 2020 relative to 2001. Source: Banse *et al.* (2008a,b).

then the impact on world prices is substantial and the long term trend of declining world prices in the reference scenario might be dampened or reversed. The arrival and impact of second generation biofuels is uncertain. According to Banse *et al.* (2008a,b), biofuels lead to higher agricultural income, land use and land prices, and a loss of biodiversity.

Development of oil prices is crucial for the development of biofuels. Some experts point that prices stay high due to increased demand in Asia and depleting supply resources. Others indicate that this is a temporary situation as capacity is lacking at the moment due to too few investments in the past. If oil prices stay high, food and energy markets will be more interlinked. The oil prices will then put both a floor and a ceiling[18] for prices in the food markets (Schmidhuber, 2007). As energy markets are more elastic, the long-term trend of food prices might be changed (less negative to positive dependent on development oil price).

- High feedstock prices make biofuels less profitable (ceiling effect), as does a low oil price (floor effect). Even at current level of crude oil prices of 120 USD per barrel almost no biofuels are economically viable without policies. A low oil price implies that only biofuels will be produced under mandates or that they are heavily subsidized. Without an increase in oil prices the impact of biofuels is therefore limited to the impact of filling the mandates.

[18] Ceiling price effect: as feedstock costs are the most important cost element of all (large scale) forms of bioenergy use, feed stock prices (food and agricultural prices) cannot rise faster than energy prices in order for agriculture to remain competitive in energy markets. Floor price effect: if demand is particular pronounced as in the case of cane-based ethanol, bioenergy demand has created a quasi intervention system and an effective floor price for sugar in this case.

- The interrelation with the energy markets may slowdown or reverse Cochrane's treadmill or Owens development squeeze which imply declining real agricultural prices, less farmers, larger scale farming and possible depopulated areas.
- Volatility of world prices might be an important problem in the future that causes hunger in terms of very high prices for poor consumers and problems for poor farmers when prices are low. The ceiling and especially the floor may act as an intervention price in case of very volatile prices. A floor may also stimulate agriculture in the (poor) world. Hunger is not a problem directly related with biofuels but often of bad policies, and improperly functioning factor and commodity markets.[19] In principle, there is enough food in the world but there is a distribution problem.
- Rising food commodity prices tend to negatively affect lower income consumers more than higher income consumers. First, lower income consumers spend a larger share of their income on food. Second, staple food commodities such as maize, wheat, rice, and soybeans account for a larger share of food expenditures in low-income families. Third, consumers in low-income, food-deficit countries are vulnerable because they must rely on imported supplies, usually purchased at higher world prices. Fourth, countries receiving food aid donations based on fixed budgets receive smaller quantities of food aid. A simplified comparison of the impact of higher food commodity prices on consumers in high-income countries and on consumers in low-income, food-deficit countries illustrates these differences (see Table 2).

[19] AG assessment (2008), 'Policy options for improving livelihoods include access to microcredit and other financial services; legal frameworks that ensure access and tenure to resources and land; recourse to fair conflict resolution; and progressive evolution and proactive engagement in Intellectual Property Rights (IPR) regimes and related instruments.'

Table 2. Impact of higher food commodity prices on consumers' food budgets.

	High income countries	**Low income, food deficit countries**
Initial situation		
Income	€ 40,000	€ 1,000
Food expenditure	€ 4,000	€ 500
Food costs as % of income	10%	50%
30% increase in food prices		
New costs for total food expenditure	€ 5,200	€ 650
Food costs as % of income	13%	65%

This illustrative comparison shows that for a consumer in a high-income country a 30-percent increase in food prices causes food expenditures to rise 3 percent (€1,200), while for a consumer in a low-income country food expenditures increase by 15 percentage points.

6. Concluding remarks

The motivation at the origin of this chapter can be summarised in four questions:

- Is the current price increase driven by real or monetary issues (notably a speculation phenomenon)?
- Are natural resource and basic food commodity prices linked together?
- Is the shortfall in production also linked to governance issues that limit investment and production?
- To what extent is the underused capacity in land and man-power a result of lack of investment capacity, both at the micro level (tools and seed) and at the macro level (storage and transportation infrastructure)?

The work on these questions allows the formulation of responses, and also some broader observations. From our work it is clear that the price increases have several roots and that a normally functioning market will in time provide a certain degree of corrective action. But policy/political decisions can prevent the market from doing so. In any case, the time lapse for the market to act does not remove the acuity of the price distortion that affects the poorest people, and urgent intervention is necessary to alleviate the effects of short-term price peaks.

Natural resource prices *lead* basic food commodity prices; the rate of growth of the former has historically been (and is again at present) higher than the latter. Biofuels create a more direct link between food and fuel prices, if fuel prices are high: the long-term trend of declining real food prices might be dampened or reversed.

The influence of policy/political decisions mentioned above is certainly present when considering why production in many countries is below the potential capacity to produce food. Not only has land been voluntarily removed from production in some cases, but the access to technology and markets is sometimes also limited by factors that are strictly in the realm of governance. But then there are also potential producers, who simply can not make it into the market, and they can be assisted through micro-credit or through the donation of tools, seeds and the development of irrigation, storage capacity and transportation facilities to integrate into market structures.

Our further observations are of several orders, and theses are with regard to policy implications, market failure, social equity, and required policy action.

6.1. Policy implications

With regard to the EU, CAP reform was designed to enforce farmers' reaction to market signals. There should be no surprise, therefore, when farmers do, and therefore production falls close to the level of world demand. The problem, however, is the time lag between the demand in the market and a farmer's decision on what – and how much – to plant. There is always some degree of 'inadequate' response on the supply side. Around the world, farmers are now responding to price signals and are increasing their production of cereals. Building up and managing stocks is not the primary responsibility of farmers, and in a free market this is left to traders; some government intervention might be considered, but a return to automatic intervention based solely on commodity prices should be absolutely avoided!

6.2. Will current price level persist?

High prices can only 'cured' by high prices. This may initially seem to be a provocative statement, but the simple fact is that – as stated above – farmers do react to price signals. So do all the other agents in the economy, including speculators! The food price 'crisis' will certainly be prolonged through protective measures by national governments, although the issue of civil stability may encourage some governments to take such actions, to reassure their populations that 'something is being done'. Biofuels, however, create a more direct link between food and fuel prices and if fuel prices increase further, the long-term trend of declining real food prices might be dampened or reversed.

6.3. Who is mostly affected?

The consumers of food in low-income countries with food and energy deficits are those who will suffer most in any sudden or rapid price shift for basic commodities, of which foremost is food. In principle, current high prices provide additional income opportunities for farmers. Whether farmers in developing countries will benefit from current high prices on world food markets remains questionable and depends on the degree of integration of regional in global food markets. But if there is no structural market failure involved *per se*, as stated above, then this means that the conditions of productivity and market access are the priorities that have not been addressed successfully for a long period of time *before* a price crisis occurs.

6.4. Required policy action

Short-term action is to urgently increase spending on food aid (which has gone down during the last years). Long-term production capacity improvement (including publically financed agricultural research) is essential to avoid repeated price crises. The current crisis is not a crisis in terms of shortage of food, but a crisis in terms of income shortage (in terms of purchasing power and of investment potential to increase productive capacity). Policy measures should enable especially the poor to be able to participate in the economy, and therefore for the poor countries to generate income within a world market.

Acknowledgements

We consulted the following experts: Patt Westhoff (FAPRI), Josef Schmidhuber (FAO), Loek Boonekamp (OECD), Ron Trostle (ERS/USDA), Pavel Vavra (OECD), Willie Meyers (FAPRI) and Pierre Charlebois (Agriculture and Agri-Food Canada). Furthermore, we benefited greatly from insights and the discussions during the World Agricultural Outlook Conference, organized by ERS/USDA Washington DC, May, 14-15, 2008 and the Modeling Workshop on Biofuel, May, 16 organized by Farm Foundation and ERS/USDA Washington DC.

References

Banse, M., H. van Meijl, A. Tabeau and G. Woltjer, 2008a. Will EU Biofuel Policies affect Global Agricultural Markets? European Review of Agricultural Economics, 35: 117-141.

Banse, M., H. van Meijl and G. Woltjer, 2008b. The Impact of First Generation Biofuels on Global Agricultural Production, Trade and Land Use. Paper presented at the 12th EAAE Conference. 26-29 August 2008, Gent, Belgium.

Licht, F.O., 2007. Licht Interactive Data. Available at: http://statistics.fo-licht.com/. Accessed 20/04/2007.

Leibtag, E., 2008. Corn Prices Near Record High, But What About Food Costs? Amber Waves.

International Monetary Fund (IMF), 2008. International financial statistics. Available at: http://www.imfstatistics.org/imf/. Accessed 18/04/2008.

Nowicki, P., H. van Meijl, A. Knierim, M. Banse, J. Helming, O. Margraf, B. Matzdorf, R. Mnatsakanian, M. Reutter, I. Terluin, K. Overmars, D. Verhoog, C. Weeger and H. Westhoek, 2006. Scenar 2020 - Scenario study on agriculture and the rural world. Contract No. 30 - CE - 0040087/00-08. European Commission, Directorate-General, Belgium.

OECD-FAO, 2008. Agricultural Outlook 2008-2017. Paris, France.

Pardey, P.G., N. Beintema, S. Dehmer and S. Wood, 2006. Agricultural research: a growing global divide, IFPRI, Washington, USA. DOI: http://dx.doi.org/10.2499/089629529X.

Schmidhuber, J., 2007. Biofuels: An Emerging Threat to Europe's Food Security? Impact of an Increased Biomass Use on Agricultural Markets, Prices and Food Security: A Longer-term Perspective. Available at: www.notre-europe.eu.

Trostle, R., 2008. Global Agricultural Supply and Demand: Factors Contributing to the Recent Increase in Food Commodity Prices. ERS/USDA. WRS-0801.

United States Department on Agriculture (USDA), 2008. PSD view database. Available at: http://www.fas.usda.gov/psdonline/. Accessed 21/05/2008.

United States Department on Agriculture (USDA), 2008. USDA Agricultural Projections to 2017. Long-term Projections Report. OCE-2008-1. February 2008. Washington D.C.

Van Meijl, H., T.J. Achterbosch, A.J. de Kleijn, A.A. Tabeau and M. Kornelis, 2003. Prijzen op agrarische wereldmarkten; Een verkenning van projecties [Agricultural world market prices; an explorative study to projections]. Agricultural Economics Research Institute, Rapport 8.03.06.

Von Braun, J., A. Ahmed, K. Asenso-Okyere, S. Fan, A. Gulati, J. Hoddinott, R. Pandya-Lorch, M.W. Rosegrant, M. Ruel, M. Torero, T. van Rheenen and K. von Grebmer, 2008. High Food Prices: The What, Who, and How of Proposed Policy Actions. IFPRI Policy Brief, Washington, USA.

World Bank, 2008. World Bank data base. Available at: http://web.worldbank.org/ Accessed 18/04/2008.

World Bank, 2008. World Development Report 2008: Agriculture for Development. Washington DC.

Acknowledgements

We would like to express our gratitude to the authors of the different chapters of this book for their valuable contributions, their creativity, positive attitude and willingness to accept our role of coordinators. The thorough work of the peer reviewers contributed to the quality of the contents of the different chapters.

The choice of the subjects, the selection of authors, the selection of peer reviewers and the structure of the publication was ours.

Regarding Chapter 10, the underlying study - Why are current food prices so high? by Martin Banse, Peter Nowicki and Hans van Meijl, The Hague, 2008 - was financed by the Ministry of Agriculture, Nature and Food Quality, for which they are expressing their thanks as well.

We thank Wageningen International, part of Wageningen University and Research Centre to facilitate the management of the project. Finally, we gratefully acknowledge the important contribution of Wageningen Academic Publishers. Mike Jacobs and his team did a great job within a limited timeframe.

Peter Zuurbier
Jos van de Vooren

Authors

Dr. Marcos **Adami**, senior researcher at INPE - Instituto Nacional de Pesquisas Espaciais, Divisão de Sensoriamento Remoto, Jose dos Campos (SP), Brazil.

Daniel **Alves de Aquiar**, Msc and currently researcher at INPE - Instituto Nacional de Pesquisas Espaciais, Direção, Coordenação Geral de Observação da Terra, Jose dos Campos (SP), Brazil.

Dr. Weber Antonio Neves do **Amaral** is full professor at the University of São Paulo, ESALQ - Escola Superior de Agricultura 'Luiz de Queiroz', Piracicaba (SP), Brazil.

Laura Barcellos **Antoniazzi**, Msc is working at ICONE - Instituto de Estudos do Comércio e Negociações, São Paulo (SP), Brazil.

Dr. Miriam Rumenos Piedade **Bacchi**, researcher at CEPEA - Centro de Estudos Avançados em Economia Aplicada, University of São Paulo, ESALQ - Escola Superior de Agricultura 'Luiz de Queiroz', Piracicaba (SP), Brazil.

Dr. Martin **Banse**, senior researcher at the Agricultural Economics Research Institute of Wageningen University and Research Centre, The Hague, the Netherlands.

Dr. Augusto **Beber**, researcher at Venture Partners do Brazil - São Paulo (SP), Brazil.

Dr. Annie **Dufey**, senior researcher at IIED - International Institute for Environment and Development, London, United Kingdom.

Dr. Andre **Faay**, associate professor at the Copernicus Institute, Utrecht University, the Netherlands.

Dr. Günther **Fischer** leads the Land Use Change and Agriculture Program (LUC) at IIASA - International Institute for Applied Systems Analysis, in Laxenburg, Austria.

Dr. Eduardo **Giuliani** is partner at Venture Partners do Brazil - São Paulo (SP), Brazil

Dr. Eva Tothne **Hizsnyik** joined the Land Use Change and Agriculture Program (LUC) at IIASA - International Institute for Applied Systems Analysis - in 2003 as a part-time Research Scholar, Laxenburg, Austria.

Dr. Isaias **Macedo** is visiting researcher at NIPE - Núcleo Interdisciplinar de Planejamento Energético, Universidade Estadual de Campinas (UNICAMP), Campinas (SP), Brazil. Since 2001, he has been consultant in Energy for the Brazilian Government and the private sector.

João Paulo **Marinho**, Msc is graduate student at the University of São Paulo, ESALQ - Escola Superior de Agricultura 'Luiz de Queiroz', Piracicaba (SP), Brazil.

Dr. Andre Meloni **Nassar** is director-general ICONE - Instituto de Estudos do Comércio e Negociações, São Paulo (SP), Brazil.

Dr. Peter **Nowocki**, senior researcher at the Agricultural Economics Research Institute of Wageningen University and Research Centre, The Hague, the Netherlands.

Dr. Bernardo F.T. **Rudorff** is senior research at INPE - Instituto Nacional de Pesquisas Espaciais, Divisão de Sensoriamento Remoto, Jose dos Campos (SP), Brazil.

Dr. Joaquim E.A. **Seabra**, professor at FEM - Faculdade de Engenharia Mecânica, Universidade Estadual de Campinas (UNICAMP), Campinas (SP), Brazil.

Dr. Alfred **Szwarc**, consultant at ADS - Technology and Sustainable Development, São Paulo, Brazil.

Rudy **Tarasantchi**, Msc is graduate student at the University of São Paulo, ESALQ - Escola Superior de Agricultura 'Luiz de Queiroz', Piracicaba (SP), Brazil.

Dr. Edmar **Teixeira** joined the Land Use Change and Agriculture Program of IIASA - International Institute for Applied Systems Analysis, in April 2007 as a Postdoctoral Research Scholar, Laxenburg, Austria.

Dr. Wallace E. **Tyner**, professor at the Department of Agricultural Economics, Purdue University, West Lafayette (IN), USA.

Dr. Jos **van de Vooren** is manager of the Latin America Office Wageningen University and Research Centre, the Netherlands, located at University of São Paulo, ESALQ - Escola Superior de Agricultura 'Luiz de Queiroz', Piracicaba (SP), Brazil.

Dr. Hans **van Meijl**, senior researcher at the Agricultural Economics Research Institute of Wageningen University and Research Centre, The Hague, the Netherlands.

Dr. Harrij **van Velthuizen** is a land resources ecologist and specialist in agro-ecological zoning. Since 1995 he has been engaged with the activities of the Land Use Change and Agriculture Program (LUC) of IIASA - International Institute for Applied Systems Analysis, Laxenburg, Austria..

Dr. Arnaldo **Walter**, professor at UNICAMP - Universidade Estadual de Campinas, Campinas (SP), Brazil.

Dr. Peter **Zuurbier** is director of the Latin America Office of Wageningen University and Research Centre, the Netherlands. He is also professor at the University of São Paulo, ESALQ - Escola Superior de Agricultura 'Luiz de Queiroz', Piracicaba (SP), Brazil and at Wageningen University.

Keyword index

Printed in the United States
by Baker & Taylor Publisher Services